EARTH AND SPACE Science
Lab Manual

Colorado Springs, Colorado

© 2017 by ACSI/Purposeful Design Publications
All rights reserved.

Printed in the United States of America
26 25 24 23 22 21 20   2 3 4 5 6 7 8

Earth and Space Science – Lab Manual
Purposeful Design Science series
ISBN 978-1-58331-545-3, Catalog #20083

No portion of this book may be reproduced, stored in a retrieval system, or transmitted, in any form or by any means—mechanical, photocopying, recording, or otherwise—without prior written permission of ACSI/Purposeful Design Publications.

Purposeful Design Publications is the publishing division of the Association of Christian Schools International (ACSI) and is committed to the ministry of Christian school education, to enable Christian educators and schools worldwide to effectively prepare students for life. As the publisher of textbooks, trade books, and other educational resources within ACSI, Purposeful Design Publications strives to produce biblically sound materials that reflect Christian scholarship and stewardship and that address the identified needs of Christian schools around the world.

References to books, computer software, and other ancillary resources in this series are not endorsements by ACSI. These materials were selected to provide teachers with additional resources appropriate to the concepts being taught and to promote student understanding and enjoyment.

Unless otherwise noted, all Scripture quotations are taken from THE HOLY BIBLE, NEW INTERNATIONAL VERSION®, NIV® Copyright © 1973, 1978, 1984, 2011 by Biblica, Inc.® Used by permission. All rights reserved worldwide.

Cover Design: Mike Riester

Purposeful Design Publications
A Division of ACSI
731 Chapel Hills Drive • Colorado Springs, CO 80920
Customer Service Department: 800-367-0798
Website: www.purposefuldesign.com

Name: _____ Date: _____

Lab 1.1.2A Half-Lives

QUESTION: What do half-lives teach about Creation?

HYPOTHESIS: _____

_____

EXPERIMENT:

| **You will need:** | • 1 cm × 8 cm strip of paper | • scissors |

Steps:
1. Create a tally sheet to record the number of cuts made in Steps 2 and 3.
2. The strip of paper represents the carbon 14 in a sample. Cut the paper in half. Discard half of the strip. Record this cut on the tally sheet.
3. Cut the remaining half in half, and discard half of the strip. Record this cut on the tally sheet.
4. Repeat Step 3 until it is no longer possible to cut the strip in half.

ANALYZE AND CONCLUDE:

1. How many times were you able to cut the paper in half until it was too small to cut? _____

_____

2. What length of time is represented by each cut? _____

3. What length of time is represented by the total amount of cuts? _____

_____

4. Could you use the half-life of carbon 14 to prove that dinosaurs died millions of years ago? _____

_____

5. Can carbon-14 dating be used to determine the age of the earth? Why? _____

_____

_____

6. Uranium 238 has a half-life of 4.5 billion years. Explain why uranium is used instead of carbon 14 to date things older than 50,000 years old. _____

_____

_____

© Earth and Space Science • Introduction to Earth Science

Name: _____ Date: _____

Lab 1.1.3A Topographic Map

QUESTION: What can you learn from a topographic map?

HYPOTHESIS: _____
_____

EXPERIMENT:

You will need:	• metric ruler	• graph paper
• modeling clay	• tub	

Steps:
1. Make a model landscape of a mountain and hills of varying slopes with the modeling clay. Make the landscape at least 8 cm high at its highest point.
2. Carefully press the landscape into an empty tub.
3. Hold a metric ruler upright in the tub next to the landscape.
4. Pour water into the tub until the water reaches a depth of 1 cm. Hold the landscape in place and carefully trace around the water level on the clay with a sharp pencil.
5. Repeat Step 4, raising the level of the water 1 cm at a time, tracing the water level each time. Do this until water covers the landscape.
6. Carefully remove the water from the tub. Be careful not to rub off the marks. Trace over the marks again so they show up better.
7. Look straight down at the top of the landscape. On graph paper, sketch the rings as seen from above.
8. Compare the sketch to the landscape.

ANALYZE AND CONCLUDE:

1. How does the number of rings relate to the height of the mountain and hills?
_____
_____
_____

2. How does the distance between rings relate to the slope of the mountain and hills?
_____
_____
_____

© Earth and Space Science • Introduction to Earth Science

Name: _____   Date: _____

Traveling Across Time Zones WS 1.1.3A

Teams of geologists are flying to England to study the limestone in the White Cliffs of Dover. All teams will fly into Heathrow Airport in London. Calculate their arrival time in London using their departure times, flight times, and **WS 1.1.3B World Time Zones Map**.

Travel Schedule

Departure City	Departure Time	Trip Duration	Arrival Time
Astana, Kazakhstan	11:00 AM	7 hours	
La Paz, Bolivia	6:30 AM	20 hours	
Los Angeles, United States	4:00 PM	10.5 hours	
Luanda, Angola	12:00 PM	8.25 hours	

Name: _____ Date: _____

World Time Zones Map

WS 1.1.3B

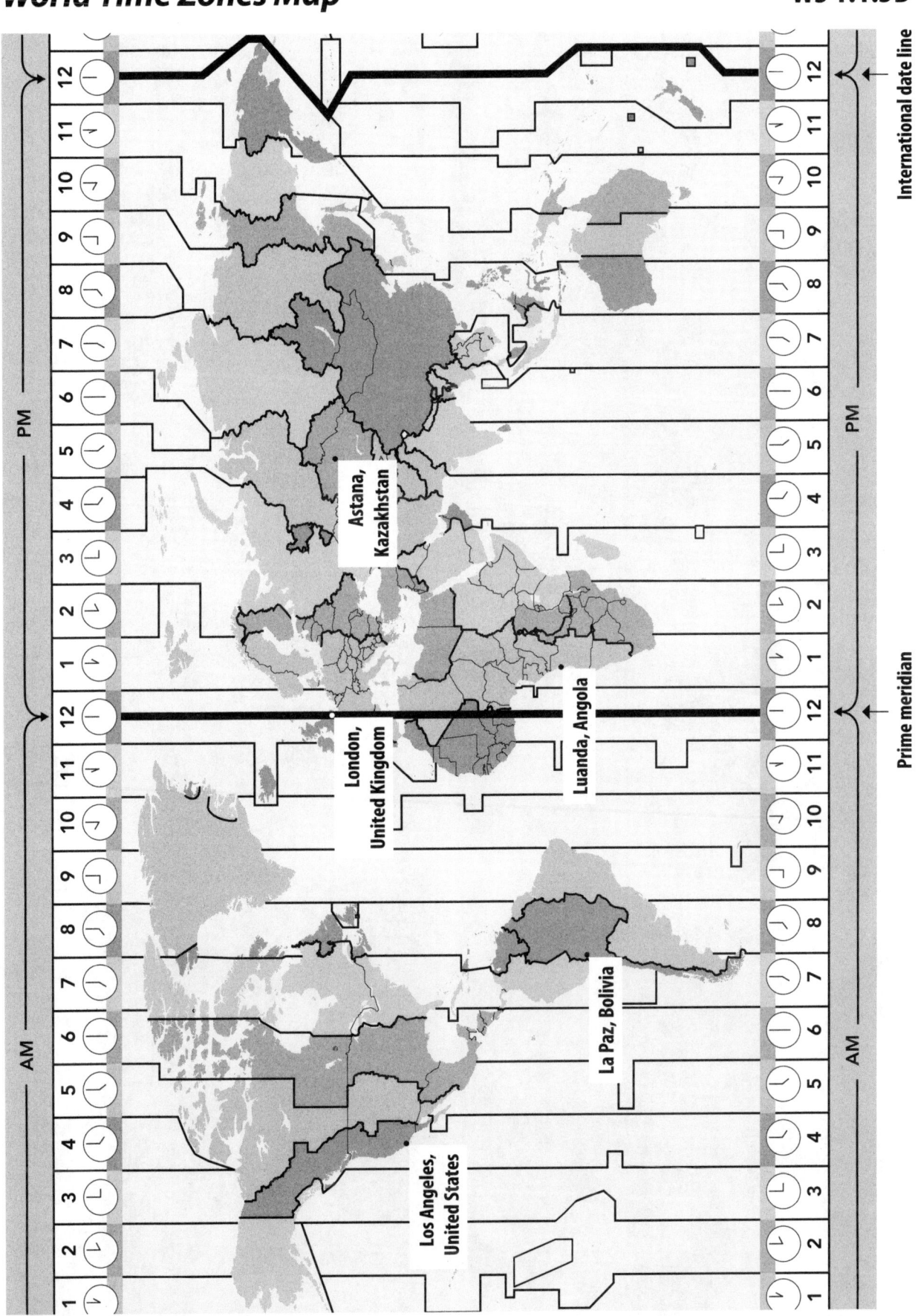

© Earth and Space Science • Introduction to Earth Science

Name: _____ Date: _____

Map Selection   WS 1.1.3C

Answer the following questions:

1. You are testing the effects of altitude (height above sea level) on baking bread. What kind of map would you use to make sure you perform your tests at different altitudes? _____

2. What kind of map would help you determine how erosion in a particular area is affected by the amount of rain and snowfall it receives? _____

3. While visiting a museum, you see the fossilized remains of a saber-toothed cat. This animal thrived for a long time before becoming extinct many years ago. You are interested in what the land looked like when this creature was alive. What kind of map should you consult?

4. Tomorrow you plan to go on a hike, and you want to know whether you will need a jacket. What kind of map will help you decide? _____

5. You are planning to build a house, but your area experiences earthquakes. What kind of map will help you avoid building too close to a fault? _____

Name: _____ Date: _____

Tracking Down the Treasures   WS 1.1.3D

Use a globe or a map with lines of latitude and longitude to discover the location of valuable minerals.

1. This rain forest country is rich in diamonds, and it can be found at 10° W longitude and 10° N latitude. Identify the country and the continent on which it is found. _____

2. This country is thickly covered with tropical forests. Some of the world's finest emeralds can be found here. Locate the South American country and its capital city, which is found on the 5° N latitude directly below the country of Haiti.
_____

3. Oil and natural gas can be found in these two Canadian provinces, both of which are sandwiched between the 50° N and 60° N latitudes. The 110° W longitude is the dividing line between them. _____

4. Search for precious opals near this capital city located below the Arctic Circle and above the Black Sea. _____

5. This country is rich in diamonds, gold, uranium, platinum, coal, and iron. It is located just below the Tropic of Capricorn, and it is bordered by the Atlantic Ocean and Indian Ocean. _____

6. Beautiful sapphires are mined in the eastern mountains of this continent in the Southern Hemisphere, which is southwest of the largest coral reef in the world. Identify both the continent and the mountains containing the sapphires.
_____

7. One of the most fabulous gem collections in the world can be found in a museum in this capital city near the intersection of 40° N latitude and 80° W longitude. _____

8. The equator cuts through this southeast Asian island, which is rich in both diamonds and biodiversity. Search directly below Hong Kong. _____

9. Iron and copper deposits are located in this bay that lies 52°–62° 50' N latitude and 76°–95° W longitude. _____

10. Titanium and zircon can be found in the waters off the shores of this large island. Search in the time zone three hours ahead of the time zone containing the prime meridian. _____

© Earth and Space Science • Introduction to Earth Science

Name: _____ Date: _____

Lab 1.2.3A Mineral Identification

QUESTION: What minerals are these?

HYPOTHESIS: _____
_____

EXPERIMENT:

You will need:	• steel file	• triple beam balance
• mineral samples	• glass plate	• graduated cylinder of water
• mineral test kit	• streak plate	

Steps:
1. Describe the specimen's color.
2. Refer to the luster chart and examples in the Student Edition to help you describe the specimen's luster.
3. Scrape the specimen firmly across the unglazed side of the streak plate. You may need to do this a couple times. If you cannot see anything, feel the streak plate for any residue. If there is residue, the streak is white or clear. If there is no residue, there is no streak.
4. Firmly wipe your fingers back and forth across the specimen. If it leaves a residue on your fingers, the hardness is a 1; go to Step 10. If it does not leave a residue, go to Step 5.
5. Try to scratch the specimen with your fingernail. If it leaves a scratch, the hardness is 2; go to Step 10. If there is no scratch, go to Step 6.
6. Use the copper coin to try to scratch the specimen. If it leaves a scratch, it has a hardness of 3; go to Step 10. If there is no scratch, go to Step 7.
7. Try to scratch the specimen with the steel nail. If it leaves a scratch, the hardness is 4–5; go to Step 10. If there is no scratch, go to Step 8.
8. Try to scratch the specimen with the steel file. If it leaves a scratch, the hardness is 6; go to Step 10. If there is no scratch, go to Step 9.
9. Use the specimen to try to scratch the piece of glass. If it scratches the glass, it has a hardness of 8. If it does not scratch the glass, the hardness is a 9–10.
10. Use the triple beam balance to find the specimen's mass.
11. Fill the beaker approximately half full with water and record the volume to the nearest milliliter. Carefully place the specimen in the beaker so as to avoid splashing. Record the volume to the nearest milliliter. Subtract the volume without the specimen from the volume with the specimen and record the difference on the data chart in cubic centimeters. Remember, 1 mL = 1 cm^3.
12. Determine the specimen's density by dividing its mass by its volume.
13. Compare your findings to the information on the Mineral Properties Chart in the Student Edition, or to other sources found online or provided by the teacher. Identify the mineral on the data chart.

Lab 1.2.3A Mineral Identification

Mineral Data Chart

	Specimen A	Specimen B	Specimen C	Specimen D
Color				
Luster				
Streak				
Hardness				
Mass (g)				
Volume (cm^3)				
Density (m/V)				
Identity				

ANALYZE AND CONCLUDE:

1. Of the minerals you tested, which had the greatest mass? _____

_____

2. Of the minerals you tested, which had the greatest volume? _____

_____

3. Of the minerals you tested, which had the greatest density? _____

_____

4. If volume remains constant but mass increases, what will happen to density?

_____

5. If mass remains constant but volume increases, what will happen to density?

_____

6. Which mineral is the hardest? _____

_____

7. Which minerals' colors approximately matched their streaks? _____

_____

8. Why are so many tests necessary when identifying minerals? _____

_____

_____

Name: _____ Date: _____

Precious Jewels   WS 1.2.2A

Revelation 21 paints an amazing picture of the New Jerusalem with walls composed of minerals and gemstones. Read Revelation 21:15–21. Use resources to complete the chart. List the gemstones mentioned in the passage and use resources to identify the following characteristics of each gem: mineral composition, chemical formula, mineral classification, and color.

Gemstones and Minerals

Gemstone	Parent Mineral	Chemical Formula	Mineral Classification	Color
Jasper				
Sapphire				
Agate				
Emerald				
Onyx				
Ruby				
Chrysolite				
Beryl				
Topaz				
Jacinth				
Amethyst				

Note: Some formulas may have two elements in parentheses separated by a comma. The comma indicates that either element may be used for this part of the equation. For example, the formula for the sulfide mineral pentlandite is $(Fe, Ni)_9S_8$; iron or nickel can bond with sulfur to form the same mineral. If no comma is present, then the parentheses show that an ion occurs multiple times in the molecule. For example, the formula for the carbonate mineral dolomite is $CaMg(CO_3)_2$; one calcium atom and one magnesium atom combine with two carbonate ions.

© Earth and Space Science • Minerals

Name: _____ Date: _____

Periodic Table of Elements

WS 1.2.2B

PERIODIC TABLE OF ELEMENTS

The physical state of elements 113, 115, and 117 is not confirmed but is expected to be solid. The physical state of element 118 is not confirmed but is expected to be gas.

© Earth and Space Science • Minerals

Name: _____ Date: _____

Periodic Table of Elements   WS 1.2.2B

Use colored pencils to complete the instructions.
1. Color the symbols for the following elements bright blue: 35 and 80.
2. Color the symbols for the following elements white: 1, 2, 7–10, 17, 18, 36, 54, 86, and 118.
3. In the gray rectangle, color the boxes as follows: metal, red; metalloid, yellow; nonmetal, green; unknown, gray. The symbols for solid elements should be black, the symbols for gaseous elements should be white, and the symbols for liquid elements should be blue.
4. Color the symbols for all the other elements black.
5. Shade the following elements green: 1, 2, 6–10, 15–18, 34–36, 53, 54, 86, and 118.
6. Shade the following elements yellow: 5, 14, 32, 33, 51, 52, and 85.
7. Shade the following element gray: 117.
8. Shade all other elements orange.
9. Add an asterisk (*) to elements 43, 61, 93–118.
10. Place a pink border around the first four elements in column 17 (fluorine, chlorine, bromine, iodine). In the gray rectangle, color the line labeled *halogen group* pink.
11. Place a purple border around the following elements in column 18 (helium, neon, argon, krypton, xenon, radon). In the gray rectangle, color the line labeled *noble gases group* purple.

Earth and Space Science • Minerals

Name: _____ Date: _____

Chemical Classification   WS 1.2.2C

Identify each mineral's group using its chemical formula.

Chemical Classification of Minerals

Mineral	Chemical Formula	Mineral Group
Anhydrite	$CaSO_4$	
Calcite	$CaCO_3$	
Copper	Cu	
Corundum	Al_2O_3	
Diamond	C	
Dolomite	$CaMg(CO_3)_2$	
Fluorite	CaF_2	
Galena	PbS	
Garnet	$X_3Y_2(SiO_4)_3$ (X = Ca, Mg, Fe, Mn; Y = Fe, Al, Cr, Ti)	
Graphite	C	
Halite	$NaCl$	
Hematite	Fe_2O_3	
Magnetite	Fe_3O_4	
Pyrite	FeS_2	
Sphalerite	ZnS	

© Earth and Space Science • Minerals

Name: _____   Date: _____

Special Mineral Properties

WS 1.2.4A

Perform the tests listed below and record your observations for each mineral sample provided. Use the resources provided or the Student Edition to identify each sample.

Test	Observations
Texture: Feel each side of the sample.	
Magnetism: Place the sample at the zero end of the tape measure; place the nail far enough from the sample so it does not move; slide the nail toward the sample 1 cm at a time; record the measurement at which the nail moves on its own.	
Reactivity: Place one drop of hydrochloric acid on the sample.	
Fluorescence: Place the sample in the box with a peephole and a black light; turn on the light.	
Phosphorescence: Turn off the black light but do not remove the sample from the box.	
Refraction: With the sample still in the box, shine a laser pointer at it.	
Mineral Identity:	

© Earth and Space Science • Minerals

Name: _____ Date: _____

Lab 1.3.1A Formation of Igneous Rock

QUESTION: How do the crystals in igneous rock form on the basis of their cooling pattern?

HYPOTHESIS: _____

_____

EXPERIMENT:

You will need:	• hot plate	• thermometer
• candy molds	• pure maple syrup	• hand lens
• nonstick cooking spray	• pan	

Steps:
1. Spray the candy molds with nonstick cooking spray.
2. Using the hot plate, heat pure maple syrup in a pan to the "hard-crack" stage (about 150°C). Use the thermometer to find the correct temperature.
3. Before the syrup starts to crystallize, quickly pour some syrup into a few molds.
4. Do not return pan to the hot plate. When the remaining syrup begins to crystallize, quickly pour some of it into several other molds.
5. Return syrup to hot plate and heat for a moment longer. Quickly pour the rest of the syrup into the remaining molds.
6. Allow the candy to cool.

ANALYZE AND CONCLUDE:

1. With the hand lens, examine the candy that you poured before the syrup began to crystallize. Record your observations. _____

_____

2. Examine the candy that you poured as the syrup began to crystallize. Record your observations. _____

_____

3. Examine the candy that was reheated and poured after the syrup began to crystallize. Record your observations. _____

_____

4. Explain how the formation of crystals in maple syrup compares to the formation of crystals in igneous rock. Include the duration of the cooling process in your discussion. _____

_____

© Earth and Space Science • Rocks

Name: _____ Date: _____

Lab 1.3.1B Crystal Size of Igneous Rock

QUESTION: What size crystals form in igneous rock when magma cools slowly? Quickly?

HYPOTHESIS: _____
_____

EXPERIMENT:

You will need:	• concentrated iodine solution	• match
• 2 microscope slides	• microscope	
• eyedropper or pipette	• stopwatch	

Steps:
1. Place one drop of iodine on each microscope slide.
2. Using a stopwatch, record the time it takes for iodine crystals to form at room temperature.

    Crystal formation time at room temperature: _____

3. Place the slide under the microscope and sketch your observations as the crystals form.

4. Use the match to gently heat the underside of the second microscope slide. The heat will cause the iodine crystals to form rapidly. Record the time it takes for the iodine crystals to form with heat applied.

    Crystal formation time with added heat: _____

5. Place the slide under the microscope and sketch your observations of the crystals that have formed.

© *Earth and Space Science* • Rocks

Lab 1.3.1B Crystal Size of Igneous Rock

ANALYZE AND CONCLUDE:

1. What represents magma in this laboratory experiment? _____

_____

2. Which microscope slide best represents extrusive rock formation? _____

_____

3. Which microscope slide best represents intrusive rock formation? _____

_____

4. Why do you think that magma inside the earth's crust cools slower than magma on the earth's surface? _____

_____

_____

_____

Name: _____ Date: _____

Lab 1.3.2A Take a Closer Look

QUESTION: How is sedimentary rock identified or classified?

HYPOTHESIS: _____

_____

EXPERIMENT:

You will need:	• ruler	• vinegar or dilute hydrochloric acid
• sedimentary rock samples	• eyedropper or pipette	(0.05 M)
• hand lens	• paper towels	

Steps:
Complete the charts by answering the questions below.
1. On Charts 1 and 2, list the sedimentary rock samples that have been provided.
2. Examine the grain size for each clastic sedimentary rock using a hand lens. Identify the grains that you believe make up your clastic sedimentary rock samples (round, angular, flat; gravel size [>2 mm], sand size [0.063 mm–2 mm], silt and clay size [not visible–<0.063 mm]).
3. Which samples show the best evidence of stratification (layering)?
4. Place 1 drop of water on each specimen. Which are porous? (If the water begins to seep into the rock, it is porous.)
5. Dip each of the samples in water. Which ones exhibit an earthy smell?
6. Dry the samples with paper towels. Then test to see whether the sample is a carbonate sedimentary rock or has calcium carbonate for its matrix by placing a drop of weak acid on the rock and seeing whether it will fizz. (Note: If there is a sandstone, shale, or conglomerate that fizzes, it is probably because there is the presence of a calcium carbonate cement. This reaction does not mean that the rock is a limestone; it means that the cementing agent is made of calcium carbonate.) Rinse off the samples.
7. Examine the conglomerate and breccia rock samples. What color are the matrices? Note the wide range of fragment size in each. Are the fragments interlocking or separate in each?

Lab 1.3.2A Take a Closer Look

Chart 1

Sample	Grain Size and Shape	Best Stratification	Porous?	Earthy Smell?	Reacts with Weak Acid?

Chart 2

Sample	Classification (Clastic, Chemical, or Carbonate?)	Matrix Color? (May Be n/a)	Fragments: Interlocking or Separate? (I, S or n/a)

ANALYZE AND CONCLUDE:

1. What do geologists look for in sedimentary rocks in order to identify them? _____

2. What does grain size tell geologists? _____

3. What does the acid test indicate? _____

4. What are the two steps that sediments undergo to become sedimentary rock? _____

Name: _____ Date: _____

Lab 1.3.2B Formation of Sandstone

QUESTION: How does heat, pressure, and time affect the formation of sandstone?

HYPOTHESIS: _____

_____

EXPERIMENT:

You will need:	• large paper cup	• 125 mL of water
• 150 mL of sand	• spoon	• small shell or bone
• 125 mL of plaster of paris	• 250 mL beaker	

Steps:
1. Place sand and plaster of paris in a paper cup and mix thoroughly with a spoon.
2. Pour 100 mL of water into the cup and stir the mixture for several minutes. If the mixture is too thick to stir, add a little more water. If too much water is added and the mixture becomes soupy, use the tip of a pencil or a pin to punch a very small hole near the bottom of the cup to drain the excess water.
3. Place a small shell or bone in the center of the mixture. (This is intended to simulate a fossil when the sandstone is split in half.)
4. Let the mixture dry overnight in an area where it will not be disturbed.
5. The next day, carefully peel away the paper cup to reveal a pillar of sandstone.
6. Cut the sandstone in half to reveal the "fossil."

ANALYZE AND CONCLUDE:

1. How did the "rock" form? _____

_____

2. Under what conditions might the rock have formed faster? _____

_____

3. Compare the time it took the mixture to set as sedimentary rock with the natural formation of sandstone. _____

_____

© Earth and Space Science • Rocks

Name: _____ Date: _____

Lab 1.3.3A Gneiss Foliation

QUESTION: How do a rolling pin and clay represent the formation of metamorphic rock? What conditions will cause an igneous rock to change into a metamorphic rock?

HYPOTHESIS: _____
_____

EXPERIMENT:

You will need:	• modeling clay	• 25 cm string
• gneiss sample	• colored sequins	• 2 wooden blocks

Steps:
1. Sketch the gneiss sample below. Make sure the sketch includes the mineral grain arrangement.

2. Pour the sequins onto the work area and roll the ball of clay over the sequins. Once a large number of sequins stick to the outside of the ball, knead the ball until the sequins are thoroughly mixed throughout the clay.
3. Form the clay into a ball.
4. Cut the ball in half with the string and draw a picture of your observations of how the sequins are arranged below.

© Earth and Space Science • Rocks

Lab 1.3.3A Gneiss Foliation

5. Form the clay into a ball again and place it on one end of one of the wooden blocks.
6. With the second block, slowly smear the clay ball across the surface of the first block and draw your observations below.

ANALYZE AND CONCLUDE:

1. What changed the arrangement of the sequins? _____

2. What does the arrangement of the sequins tell you about the rock's history? _____

3. What features did the gneiss sample and the smeared clay containing sequins have in common? _____

Name: _____ Date: _____

Lab 1.3.4A How Rocks Change

QUESTION: How do the characteristics of sedimentary and igneous rock compare to metamorphic rock?

HYPOTHESIS: _____
_____

EXPERIMENT:

You will need:	• sample of gneiss, granite, limestone, marble, quartzite, sandstone, shale, and slate	• 100 mL graduated cylinder or beaker
• hand lens		
• triple beam balance or electronic scale		

Steps:
1. Examine each rock sample using the hand lens and record your observations in the table.
2. Determine a method to measure the mass and volume of a rock sample.
3. Calculate the density of each rock sample and record it in the table. Remember: Density = mass ÷ volume

Sample	Rock Type (I, S, or M)	Physical Characteristics	Mass (g)	Volume (mL)	Density (g/mL)
Gneiss					
Granite					
Limestone					
Marble					
Quartzite					
Sandstone					
Shale					
Slate					

© Earth and Space Science • Rocks

Lab 1.3.4A How Rocks Change

ANALYZE AND CONCLUDE:

1. Describe how sandstone and quartzite are similar and different. _____

2. Describe how the grain size of sandstone changes during metamorphism. _____

3. Describe the textual differences observed between shale and slate. _____

4. Compare your calculated densities to those calculated by other students. Infer why the values may differ. _____

5. Evaluate the changes in density between shale and slate, sandstone and quartzite, limestone and marble, and granite and gneiss. Does density always change? Explain. _____

Name: _____   Date: _____

Igneous Rock Scavenger Hunt   WS 1.3.1A

Rocks are all around. People utilize these rocks to create buildings, decorate jewelry, form roads, and perform many other uses. Examine the rock samples found near your home or school.

1. Draw a picture of the details seen in each rock. Use a magnifying lens to look for different types of materials within the same rock.

   Rock Sample 1   **Rock Sample 2**   **Rock Sample 3**

2. Describe the characteristics of each rock. What color are the rock samples? What is the size and weight of each rock sample? What shape are the rocks? Compare your drawings and description with photos, drawings, and descriptions in a rocks and minerals field guide.

   Rock Sample 1: _____

   _____

   Rock Sample 2: _____

   _____

   Rock Sample 3: _____

   _____

3. Classify the rocks.

Texture	Felsic or Mafic	Intrusive or Extrusive	Identity
1.			
2.			
3.			

4. Decide whether the rocks brought in are mixtures. If so, infer what these mixtures might contain.

	Mixtures (Yes/No)	Mixture Composition
Rock Sample 1		
Rock Sample 2		
Rock Sample 3		

© Earth and Space Science • Rocks

Name: _____ Date: _____

Identification of Igneous Rock   WS 1.3.1B

Identify the rock samples provided.

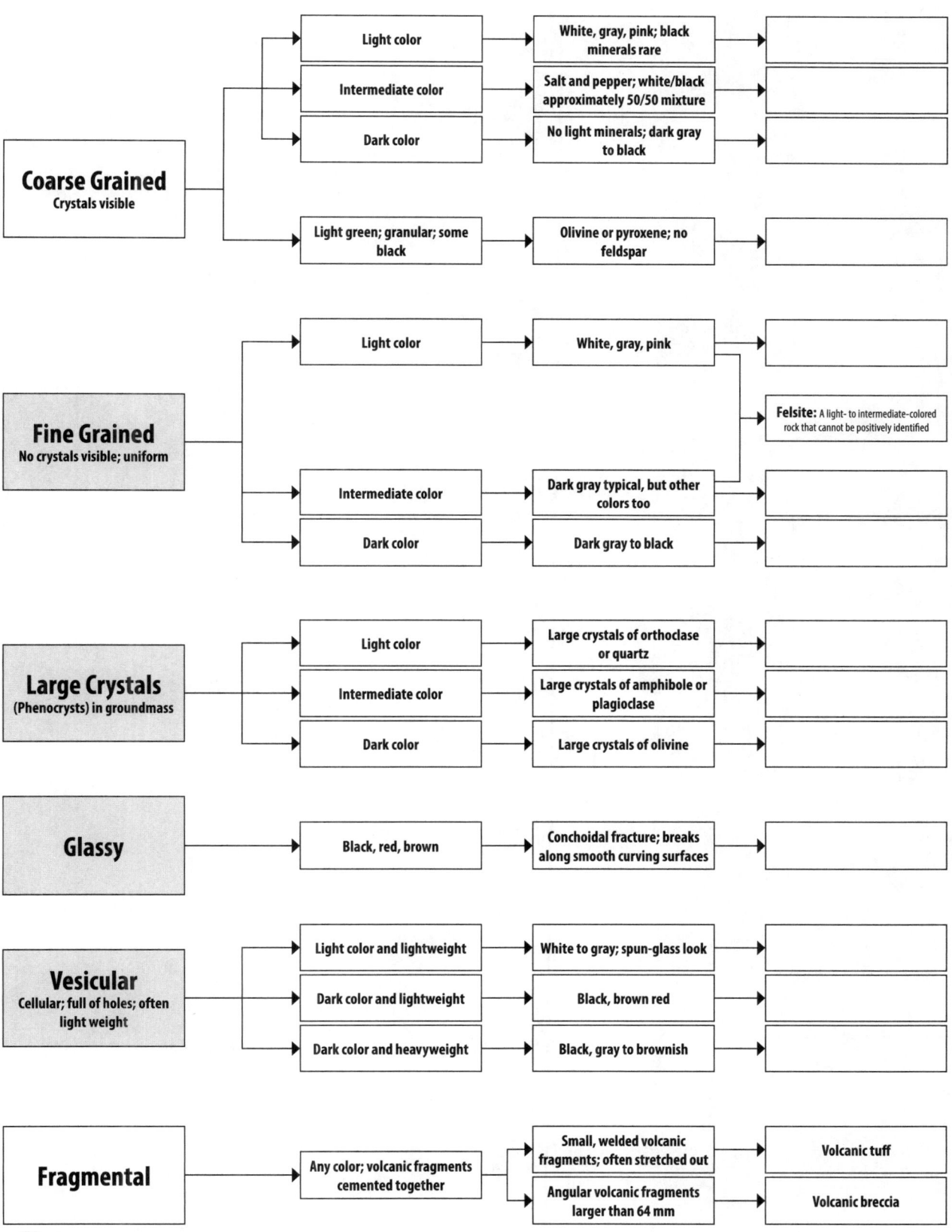

Coarse Grained — Crystals visible
- Light color → White, gray, pink; black minerals rare → ☐
- Intermediate color → Salt and pepper; white/black approximately 50/50 mixture → ☐
- Dark color → No light minerals; dark gray to black → ☐
- Light green; granular; some black → Olivine or pyroxene; no feldspar → ☐

Fine Grained — No crystals visible; uniform
- Light color → White, gray, pink → ☐
- Intermediate color → Dark gray typical, but other colors too → ☐
- Dark color → Dark gray to black → ☐

Felsite: A light- to intermediate-colored rock that cannot be positively identified

Large Crystals (Phenocrysts) in groundmass
- Light color → Large crystals of orthoclase or quartz → ☐
- Intermediate color → Large crystals of amphibole or plagioclase → ☐
- Dark color → Large crystals of olivine → ☐

Glassy
- Black, red, brown → Conchoidal fracture; breaks along smooth curving surfaces → ☐

Vesicular — Cellular; full of holes; often light weight
- Light color and lightweight → White to gray; spun-glass look → ☐
- Dark color and lightweight → Black, brown red → ☐
- Dark color and heavyweight → Black, gray to brownish → ☐

Fragmental
- Any color; volcanic fragments cemented together
  - Small, welded volcanic fragments; often stretched out → **Volcanic tuff**
  - Angular volcanic fragments larger than 64 mm → **Volcanic breccia**

© Earth and Space Science • Rocks

Name: _____   Date: _____

Sedimentary Rock Scavenger Hunt   WS 1.3.2A

Most of the visible rocks on Earth are sedimentary rock. Geologists investigate these rocks in an attempt to explain God's design of the world. Examine rocks found near your home or school to complete this worksheet.

1. Draw a picture of the details seen in each rock. Use a magnifying lens to look for different types of materials within the same rock.

   Rock Sample 1   **Rock Sample 2**   **Rock Sample 3**

2. Describe the characteristics of each rock. What color are the rock samples? What is the size and weight of each rock sample? What shape are the rocks? Compare your drawings and description with photos, drawings, and descriptions in a rocks and minerals field guide.

   Rock Sample 1: _____
   _____

   Rock Sample 2: _____
   _____

   Rock Sample 3: _____
   _____

3. Classify the rocks using the chart below.

	Sediment Size	Sediment Type	Clastic, Chemical, or Carbonate	Identity
1.				
2.				
3.				

© Earth and Space Science • Rocks

Name: _____ Date: _____

Metamorphic Rock Classification

WS 1.3.3A

Examine the metamorphic rock samples. Using a rocks and minerals field guide, classify and describe each rock.

Sample Identity	Physical Appearance	Classification (Foliated or Nonfoliated)

© Earth and Space Science • Rocks

Name: _____ Date: _____

Metamorphic Rock Scavenger Hunt   WS 1.3.3B

Did you know rocks can change into completely different rocks? Metamorphic rock is investigated by geologists in an attempt to explain historic Earth events. Examine the rocks found near your home or school to complete this worksheet.

1. Draw a picture of the details seen in each rock. Use a magnifying lens to look for different types of materials within the same rock.

   Rock Sample 1   **Rock Sample 2**   **Rock Sample 3**

2. Describe the characteristics of each rock. What color are the rock samples? What is the size and weight of each rock sample? What shape are the rocks? Compare your drawings and description with photos, drawings, and descriptions in a rocks and minerals field guide.

   Rock Sample 1: _____

   _____

   Rock Sample 2: _____

   _____

   Rock Sample 3: _____

   _____

3. Classify the rocks using the chart below.

	Color	Fine-Grained or Coarse-Grained	Foliated or Nonfoliated	Identity
1.				
2.				
3.				

© Earth and Space Science • Rocks

Name: _____ Date: _____

The Rock Cycle

WS 1.3.4A

Fill in the blanks using the following words: *magma, igneous rock, metamorphic rock, sedimentary rock, erosion, transport, deposition, melting, crystallization, pressure,* and *temperature*. Try to fill in the blanks without using your Student Edition first. Color arrows that involve heat red, weathering and erosion green, and cooling blue.

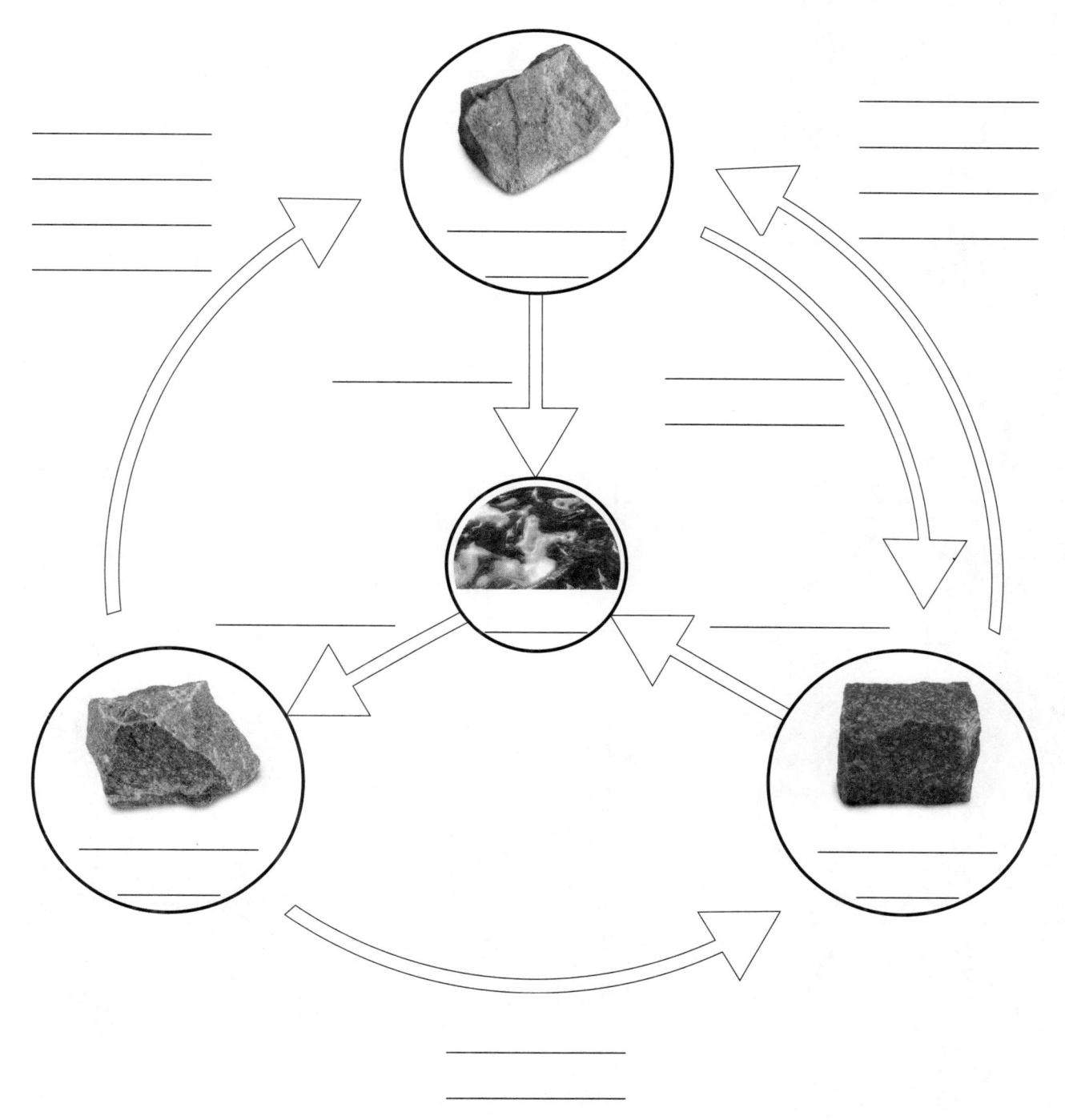

© Earth and Space Science • Rocks

Name: _____ Date: _____

Lab 1.4.1A Making an Electromagnet

QUESTION: Will an electric current in motion create a magnetic field?

HYPOTHESIS: _____
_____

EXPERIMENT:

You will need:	• insulated copper wire, 60 cm long with each end stripped, exposing the copper	• electrical or cellophane tape
• large iron nail, at least 7.5 cm long		• 1.5 volt battery (AA, C or D cell)
• paper clip		

Steps:
1. Touch the nail to the paper clip. Is there a magnetic attraction? _____
2. Tape each end of the wire to each end of the battery. Be sure the copper is making contact with the battery.
3. Tape the nail to the wire so the nail is parallel to the wire and the tip is still exposed.
4. Place the paper clip near the tip of the nail. Is there a magnetic attraction? _____
5. Remove the wire from the battery and the nail from the wire.
6. Tightly coil the wire around the nail. Be sure to leave about 3 cm at each end of the coil.
7. Tape an end of the wire to each end of the battery. Be sure the copper is making contact with the battery.
8. Place the paper clip near the tip of the nail. Is there a magnetic attraction?
_____

ANALYZE AND CONCLUDE:
1. Was there a magnetic attraction between the paper clip and the nail in Step 1?
_____

2. Was there a magnetic attraction in Step 4? _____

3. Was there a magnetic attraction in Step 8? _____

4. What caused the magnetic field in Step 8? _____
_____

5. How is this experiment like the core of the earth? _____
_____
_____
_____

© Earth and Space Science • The Structure of the Earth

Name: _____ Date: _____

Lab 1.4.2A Making Waves

QUESTION: Can indirect observations be used to draw correct conclusions?

HYPOTHESIS: _____
_____

EXPERIMENT:

You will need:	
• long hallway, preferably with a tile floor	• items that make sound as they move across the floor

Steps:
1. Divide the class into two groups—*Wavemakers* and *Listeners*.
2. Wavemakers should create a plan for making several different types of sound waves. For example, in Trial 1, a person could slowly bounce a ball toward the back of a listener. For Trial 2, a person could quickly run in a zigzag pattern toward a listener. Make plans for at least five different trials. Wavemakers should record the plan in the data table provided.
3. Listeners should set up chairs at one end of a hallway facing away from the other end.
4. Listeners should sit in the chairs. Each Listener should use a lab sheet to record his or her indirect observations in the data table provided.
5. As they are sitting in the chairs, Listeners should listen to the sounds and record the indirect observations. Each Listener should hypothesize about the characteristics of the sound.
6. Repeat Step 6 until the Wavemakers have finished making all the different sounds.
7. Listeners should now ask the Wavemakers what items actually made the sounds. Record actual items in data table.

Wavemakers' Data

Trial	Speed (slow, fast)	Direction (toward, away, back and forth)	Substance (basketball, metal chair)

Lab 1.4.2A Making Waves

Listeners' Data

Trial	Speed (slow, fast)	Direction (toward, away, back and forth)	Substance (basketball, metal chair)	Actual (fill this in when all trials are complete

ANALYZE AND CONCLUDE:

1. Which items were guessed correctly? _____

2. Which items were guessed incorrectly? _____

3. Why do you think some guesses were wrong? _____
_____
_____

4. How is this lab similar to scientists using seismic waves? _____
_____
_____
_____
_____

5. How is this lab different? _____
_____
_____
_____

6. Do you think indirect observations are a good tool for scientists to use? Why? _____
_____
_____

Name: _____ Date: _____

Lab 1.4.2B Asthenosphere

QUESTION: What causes the asthenosphere to behave like both a solid and a liquid?

HYPOTHESIS: _____

_____

EXPERIMENT:

You will need:	• disposable aluminum pan, pie pan, or baking pan	• gloves
• 2 beakers	• spoons	
• cornstarch		

Steps:
1. Measure 240 mL of cornstarch and put it into the pan.
2. Add 150 mL of water to the cornstarch and stir. Stirring will be difficult at first but keep stirring until it is a white liquid. Describe the qualities of the mixture.

   _____

   _____

3. Add another 120 mL of cornstarch to the mixture and stir. As you stir, the mixture should become more solid and more difficult to stir.

   Describe the qualities of the mixture. _____

   _____

4. As you try to stir, the mixture should become very solid. If it does not, add some more cornstarch. If it does not liquefy when you stop stirring, add some water.

5. Wearing a glove, pick up some of the mixture with your hand. Describe what

   happens. _____

   _____

6. Squeeze the mixture into a ball, then relax your hand. Describe what happens.

   _____

   _____

7. Clean up everything. Wash the mixture down the sink with hot water. Continue running hot water after the mixture is flushed down the drain.

ANALYZE AND CONCLUDE:

1. Would you classify the cornstarch mixture as a solid or a liquid? Explain.

_____

_____

_____

Lab 1.4.2B Asthenosphere

2. When does the mixture behave like a solid and when does it behave like a liquid?
_____
_____

3. How is the mixture like the asthenosphere? _____
_____
_____

4. What causes the asthenosphere to behave like a solid? What causes it to behave like a liquid? _____
_____

Name: _____ Date: _____

Model of the Earth

WS 1.4.1A

Remember that volume is the amount of space occupied by an object. Use the volumes of different colors of clay to represent the volumes of each of the earth's layers. For example, if you assume that the total model of the earth will be 300 g, multiply the percent of Earth's volume of each layer by 300. This will give you the mass you need for each layer. Refer to the Student Edition for the volumes of each layer of the earth. Record the color that you will use for each layer in the table below.

1. Measure out the amount of clay for each layer using a triple beam balance to determine the correct mass of each piece of the model. Starting with the inner core, build a model of the earth's layers.

Layer	Diameter (cm)	Volume (%)	Mass in grams	Color of the clay
Inner core				
Outer core				
Mantle				
Crust				

2. Cut the model in half using fishing line.

3. Which part of the earth has the greatest volume? Does your model reflect this?
_____
_____

4. Compared to its volume, the core makes up a relatively large amount of the earth's mass. Why might this be so? _____
_____
_____
_____

© Earth and Space Science • The Structure of the Earth

Name: _____ Date: _____

Seismic Waves
WS 1.4.1B

According to the behavior of the P and S waves, identify each part of the model of the earth as solid or liquid. Label each part and describe its general characteristics. Label the *P* and *S waves*.

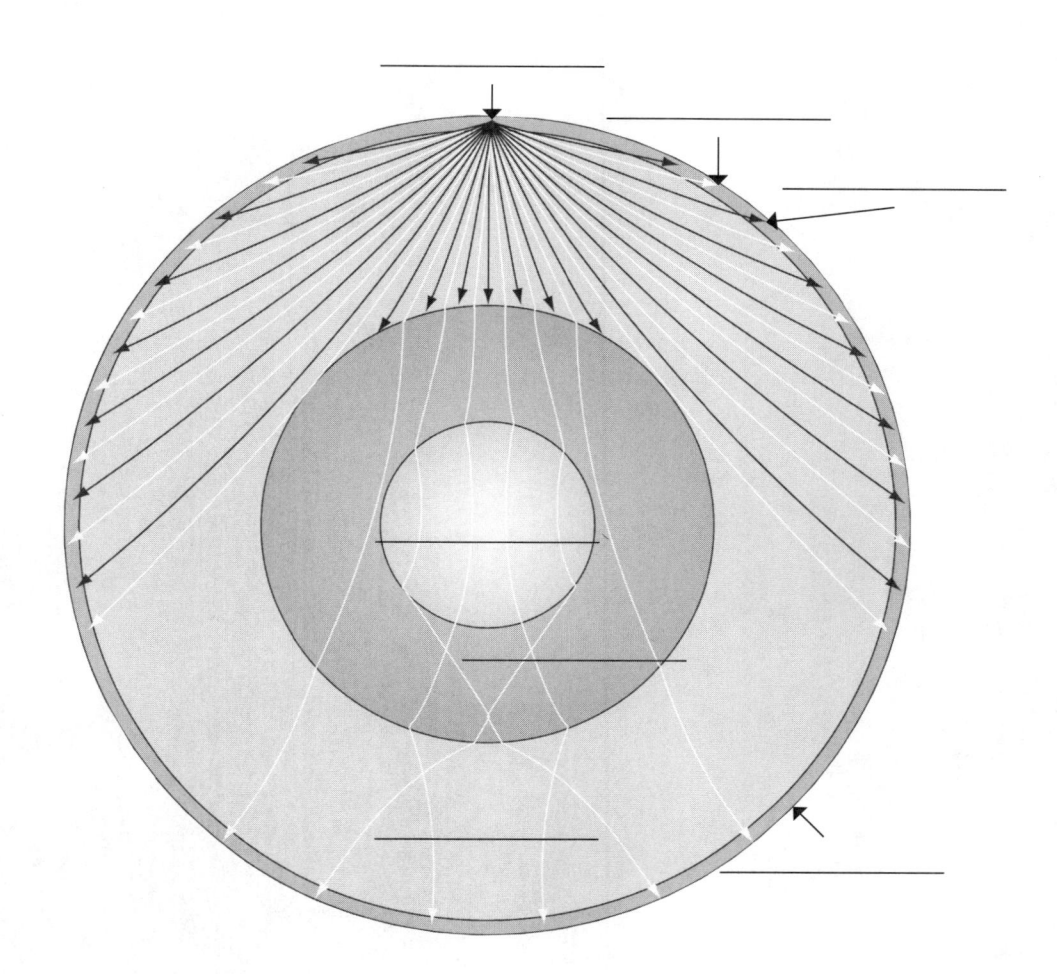

Name: _____ **Date:** _____

The Oceanic Crust   WS 1.4.3A

The oceanic crust is covered with a layer of sediments and then has separate layers of basalt.

1. Research the layers of the oceanic crust, using both the textbook and the Internet.
2. Draw a diagram showing the layer of sediments and then the two main layers of lava and dikes.
3. Include the depth, composition, and structure of each layer.

Name: _____ Date: _____

The World's Highest Mountains   WS 1.4.3B

The earth's crust has a variety of features. Each continent has mountains. Find the highest mountain peak on each continent.

1. List the name and height of the highest mountain for each continent.
2. Write the country or countries in which the mountain is located.
3. Write the latitude and longitude (DMS) coordinates for the location.
4. Write one interesting fact about the mountain.

The Highest Mountain Peaks on Each Continent

Continent	Name and Height	Country	Coordinates	Interesting Fact
North America				
South America				
Europe				
Asia				
Australia				
Africa				
Antarctica				

1. Which peak is the highest in the world? _____

2. What is the highest elevation you have been to? _____

3. What is the most interesting fact you found? _____

_____

4. Why do you think there are mountains on every continent? _____

_____

© Earth and Space Science • The Structure of the Earth

Name: _____ Date: _____

The World's Highest Mountains

WS 1.4.3B

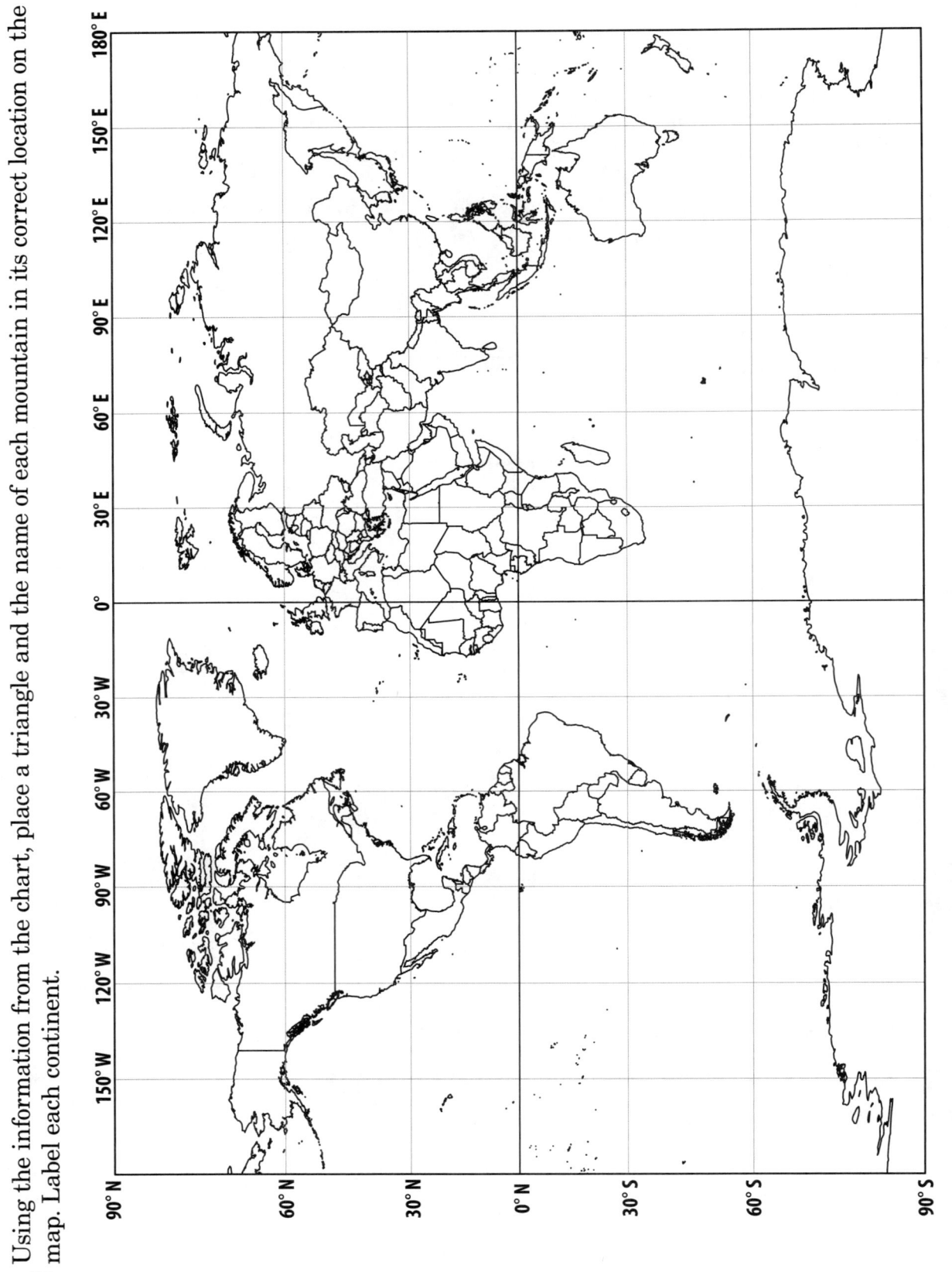

Using the information from the chart, place a triangle and the name of each mountain in its correct location on the map. Label each continent.

© Earth and Space Science • The Structure of the Earth

Name: _____ Date: _____

Diagram of Earth's Structure WS 1.4.3C

Use the illustration from the Student Edition to label each section correctly. Label the following: *inner core, outer core, mantle (lower, transition zone, upper), asthenosphere, crust,* and *lithosphere.* Use colored pencils to color each section a different color.

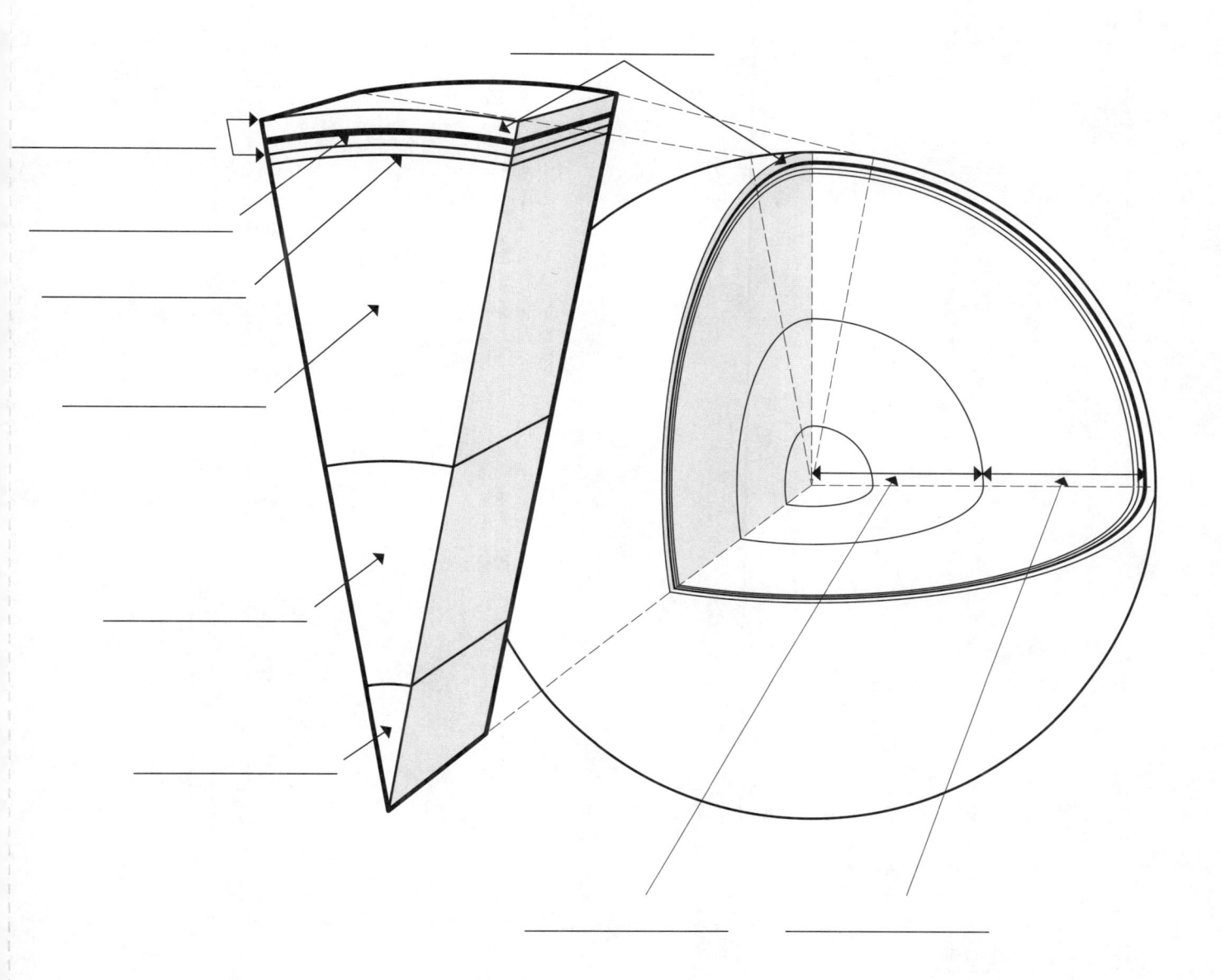

© Earth and Space Science • The Structure of the Earth

Name: _____ Date: _____

Lab 2.1.1A Ice Wedging

QUESTION: How can water break up rocks?

HYPOTHESIS: _____
_____

EXPERIMENT:

You will need:	• modeling clay	• permanent marker
• plastic drinking straw	• ziplock bag	

Steps:
1. Seal one end of a drinking straw with clay. Do not allow the clay to extend past the end of the straw.
2. Fill the straw with water.
3. Seal the other end of the drinking straw with clay. Do not allow the clay to extend past the end of the straw.
4. Place the straw inside the ziplock bag.
5. Use a permanent marker to write your name on the bag.
6. Place the straw in a freezer overnight.
7. Predict what will happen to the clay when the water in the straw freezes. Record your prediction.
8. Observe the straw the next day and record your observations. _____
_____
_____
_____
_____
_____

ANALYZE AND CONCLUDE:

1. What happened to the clay? Why did this happen? _____
_____
_____
_____

2. What do you think would happen if the water in the cracks of a rock froze? _____
_____
_____

© *Earth and Space Science* • Weathering and Erosion

Name: _____ Date: _____

Lab 2.1.2A Chemical Weathering

QUESTION: How does chemical weathering change a substance?

HYPOTHESIS: _____

_____

EXPERIMENT:

You will need:	• 2 ziplock bags	• 2 paper towels
• 2 pieces of steel wool	• permanent marker	

Steps:
1. Wet one piece of steel wool and put it in a ziplock bag. Seal the bag and use a permanent marker to label it *Wet*.
2. Place the other piece of steel wool in a plastic bag. Seal the bag and use a permanent marker to label it *Dry*.
3. Store the bags for three days.
4. Take the steel wool out of the bags and place each piece on a paper towel.

ANALYZE AND CONCLUDE:

1. Describe the appearance of the steel wool from the Dry bag. _____

_____

_____

2. Pull apart the steel wool from the Dry bag and crumble it on the paper towel.
Describe how it feels. _____

_____

_____

3. Describe the appearance of the steel wool from the Wet bag. _____

_____

4. Pull apart the steel wool from the Wet bag and crumble it on the paper towel.
Describe how it feels. _____

_____

5. How were the two pieces of steel wool different? _____

_____

_____

Lab 2.1.2A Chemical Weathering

6. How could this kind of weathering happen to a rock? _____
_____
_____

7. What kind of climates might experience this type of weathering? _____
_____
_____

8. Suggest a way to slow down or stop chemical weathering on the wet steel wool.
_____
_____
_____
_____

Name: _____ Date: _____

Lab 2.1.3A Landslide

QUESTION: How does steepness, moisture content, or an earthquake affect the stability of a slope?

HYPOTHESIS: _____

_____

EXPERIMENT:

You will need:	• 1–2 large plastic tubs	• timer
• beaker	• protractor	
• sand	• 15 mL measuring spoon	

Steps:
1. Use a beaker to measure sand and water into a large plastic tub according to the proportion given by your teacher. Thoroughly mix the sand and water together.
2. Create a hill with a 50° slope against one of the narrow ends of the tub. Verify the angle of the slope with a protractor.
3. On a separate piece of paper create a data table to record your observations.
4. Sketch your hill on a separate piece of paper.
5. Follow the steps given for your assigned experiment.

Steepness:
1. Prepare a second batch of sand and water exactly like the first one.
2. Use the second batch of sand to gradually increase the steepness of the slope by 5°–10° at a time.
3. Record the slope's steepness and describe your observations after each addition.
4. Sketch the hill, including where the landslide started and how the sand moved.

Moisture Content:
1. Use a 15 mL measuring spoon to gently pour water on the top of the hill. Be sure to pour the water the same way each time.
2. Record the amount of water added to the hill. Describe how you added the water and your observations of the hill after each water addition.
3. After a landslide occurs, tally the total amount of water added to the hill.
4. Sketch the hill, including where the landslide started and how the sand moved.

Earthquake:
1. Gently shake the tub from side to side to simulate an earthquake. Count the number of shakes needed to cause a landslide (one shake equals one side-to-side movement) and time how long the hill shakes before a landslide occurs.
2. Record the number of shakes and the time taken for a landslide to occur.
3. Calculate the shaking frequency. Divide the number of shakes by the number of seconds and label the answer as *shakes per second*. Record your answer.
4. Sketch the hill, including where the landslide started and how the sand moved.

Lab 2.1.3A Landslide

ANALYZE AND CONCLUDE:

1. How did the results of your experiment differ from similar experiments? Why might this be the case? _____
_____
_____
_____
_____
_____

2. Compare your results with groups doing the other two experiments. Which method triggered a landslide most quickly? _____
Which method caused the greatest amount of material to slide? _____
_____

3. Compare the results to your hypothesis and revise it if needed. _____
_____
_____

4. Use what you learned to suggest methods of reducing the chances of a landslide.
_____
_____
_____
_____
_____
_____
_____
_____
_____
_____

© *Earth and Space Science* • Weathering and Erosion

Name: _____ Date: _____

Lab 2.1.4A Deflation

QUESTION: How do grains erode?

HYPOTHESIS: _____

_____

EXPERIMENT:

You will need: • safety goggles • ice cube tray	• wood block • metric ruler • 300 mL of varied-grain sand	• hair dryer

Steps:
1. Put on safety goggles.
2. Lay the ice cube tray on the lab bench and place the wood block next to the tray at one end.
3. Measure the distance from the center of each section of the tray to the wood block and record these in a data table on a separate sheet of paper.
4. Form a small pile of sand on the wood block and record its height. _____
5. Hold the hair dryer close to the sand pile and record the distance between them.

   _____
6. Sketch the setup. Label all distances and the height of the sand pile.

7. Blow sand into the ice cube tray for 1 minute. Hold the hair dryer the same distance away from the sand pile at all times.
8. Use the metric ruler to measure or estimate the size of the sand grains that jumped into each section of the tray.
9. Create another data table to show the number and approximate size of the particles you found in the different sections of the ice cube tray.

© Earth and Space Science • Weathering and Erosion

Lab 2.1.4A Deflation

10. Sketch the setup again.

ANALYZE AND CONCLUDE:

1. Explain the relationship between the size of the sand grains and the distance they traveled. _____
_____
_____
_____

2. Describe how the shape of the sand pile changed. _____
_____
_____
_____

3. How might the sand pile look if the force of the blowing air was greater or if the air blew for a longer time? _____
_____
_____

Name: _____ Date: _____

Lab 2.1.4B Wave Erosion

QUESTION: How do waves affect a sandy shoreline over time?

HYPOTHESIS: _____

_____

EXPERIMENT:

You will need:	• paint roller tray	• capped, empty water bottle
• 500 mL beaker	• timer	
• sand	• metric ruler	

Steps:
1. Use 1 L of sand to create a beach in the shallow end of the paint tray. The beach should slope toward the deep end of the tray so water can cover part of it.
2. Pour 1.5 L of water into the deep end of the tray. Try not to create waves.
3. Let the sand and water sit for 5 minutes.
4. On a separate piece of paper, draw a picture of the beach.
5. Measure the width of the beach between the tub and the shoreline. If the shoreline is irregular, then take several measurements. Mark all measurements on your drawing.
6. Gently place a capped water bottle in the tray's deep end so it floats parallel to the shoreline.
7. Bob the bottle up and down for 1 minute to create waves. Try to keep the bobbing force and frequency consistent and count the number of times you bobbed the bottle.
8. Repeat Steps 3–5. Label the drawing *1 Minute*.
9. Below your drawing, describe what you observed as the bottled was bobbed. Include the approximate force of the waves (gentle, moderate, hard) and the number of times you bobbed the bottle.
10. Repeat Steps 3–5 and 9, but this time bob the bottle for 2 minutes. Try to keep the bobbing force and frequency the same as that used for the 1-minute trial. Label the drawing as *2 Minutes*.

ANALYZE AND CONCLUDE:
1. Compare the amount of erosion on the beach between the 1-minute trial and the 2-minute trial. _____

_____

_____

_____

_____

© *Earth and Space Science* • Weathering and Erosion

Lab 2.1.4B Wave Erosion

2. What happened to the sand that eroded from the shoreline? _____
_____
_____

3. How could the lab be modified to make the waves more consistent? _____
_____
_____
_____

4. Explain whether the lab would be more realistic with identical waves or with varying waves. _____
_____
_____
_____

5. Review your hypothesis. If necessary, write a more accurate answer. _____
_____
_____

Name: _____   Date: _____

Lab 2.1.5A Water Erosion and Time

QUESTION: How is gully formation affected by the amount of time a slope is exposed to runoff?

HYPOTHESIS: _____

_____

EXPERIMENT:

You will need:	• brick	• timer
• baking tray with raised edges	• protractor	• hose clamp
• sand	• hose with faucet connection	
• fine screen	• 500 mL beaker	

Steps:
1. Pack a baking tray with damp sand.
2. Place the screen over a sink drain. Place one end of the tray over the sink and place the other end on a brick.
3. Use a protractor to determine the angle of elevation for the surface of the sand.
4. On a separate piece of paper, draw and label the setup.
5. Attach the hose to the faucet and place the other end in the beaker.
6. Turn on the water so only a trickle comes out of the hose.
7. Time how long it takes to fill the 500 mL beaker.
8. Clamp the hose.
9. Calculate the rate of flow as liters per second (L/s) and record it on your drawing.
10. Move the hose to the top of the elevated pan and let the water run for 1 minute. Note how much time has passed when erosion begins.
11. Clamp the hose but do not turn off the faucet. Do not move the hose.
12. Measure, draw, and describe the erosion.
13. Allow the water to run for 1 more minute. Note how much time has passed when erosion begins.
14. Turn off the faucet.
15. Measure, draw, and describe the erosion.

ANALYZE AND CONCLUDE:

1. Compare the amount of erosion after 1 minute to the amount of erosion after 2 minutes. _____

_____

2. In the first trial, how soon were signs of erosion visible after the water started flowing over the sand? _____

3. In the second trial, how soon were additional signs of erosion visible after the water started flowing over the sand? _____

© *Earth and Space Science* • Weathering and Erosion

Name: _____ Date: _____

Lab 2.1.5B Water Erosion and Force

QUESTION: How does the speed of water affect erosion on a slope?

HYPOTHESIS: _____
_____

EXPERIMENT:

You will need:	• gravel	• hose
• shoe box	• sand	• metric ruler
• scissors	• plastic tub	• fine screen
• scale	• block	

Steps:
1. Remove the lid from a long shoe box.
2. Hold one narrow end of the shoe box facing you with the open side turned up and make a mark in the center of the end. Cut from the end's two top corners to the mark and remove the triangular piece.
3. Weigh the empty shoe box and record its mass in the data table provided.
4. Lightly pack the box with a mixture of gravel and damp sand up to the bottom of the cutout. Make the surface as flat as possible.
5. Weigh the full shoe box and record its mass in the table.
6. Subtract the mass of the empty shoe box from the mass of the full shoe box and record the mass of the material in the table.
7. Weigh the plastic tub and record its mass in the table. Place the tub in a sink.
8. Set the open edge of the shoe box on the edge of the sink so the runoff will flow into the tub.
9. Use a block to elevate the closed end of the shoe box.
10. Attach a hose to the faucet and place the other end of the hose on the upper end of the shoe box.
11. Turn on a slow stream of water and let it flow through the shoe box for 2 minutes.
12. Turn off the water.
13. Measure, draw, and describe the erosion in the shoe box.
14. Drain the contents of the tub through a screen then pour the contents of the screen back into the tub.
15. Weigh the tub and eroded material and record its mass in the table.
16. Subtract the mass of the empty tub from the mass of the full tub and record the mass of the eroded material in the table.
17. Repeat Steps 4–16, but this time use a fast stream of water.

© *Earth and Space Science* • Weathering and Erosion

Lab 2.1.5B Water Erosion and Force

	Slow Water	Fast Water
Mass of Empty Box		
Mass of Full Box		
Mass of Materials		
Mass of Empty Tub		
Mass of Full Tub		
Mass of Eroded Materials		

ANALYZE AND CONCLUDE:

1. How much more eroded material resulted from the fast water than the slow water? _____

2. Compare the dimensions of the gullies formed by the slow water and the fast water. _____

3. What differences did you observe in the erosion patterns formed by the slow and the fast water? _____

4. Compare your data table with other groups' findings. What might account for the differences you see? _____

Name: _____ Date: _____

Lab 2.1.6A Glacial Erosion

QUESTION: How is a glacier's erosion pattern affected by its environment?

HYPOTHESIS: _____
_____

EXPERIMENT:

You will need:		
• ice cup containing 20 g sand and 50 mL water	• sand	• timer
	• blocks	• thermometer
• tray	• protractor	• graph paper
	• metric ruler	

Steps:
1. Pack the tray with dry sand and use the blocks to elevate one end of the tray.
2. Use a protractor to measure the tray's angle of elevation and record it. _____
3. Get an ice cup from your teacher. Tear the paper cup off the ice.
4. Measure the height of the ice cube.
5. Place the ice cube at the top of the tray, touching the inside of the elevated edge.
6. Place a thermometer on the sand near the ice cube's anticipated path.
7. Every 5 minutes, measure the temperature as well as the ice cube's height and its distance from the tray's elevated end (measure from the inside edge of the tray). Record the measurements in the table.

	0 min	5 min	10 min	15 min	20 min	25 min	30 min	35 min	40 min
Temp.									
Ice Cube Height									
Distance Ice Cube Traveled	0.0 cm								

8. On a piece of graph paper, graph the tray's temperature in relation to time.
9. Graph the ice cube's height in relation to time.
10. Graph the distance the ice cube traveled in relation to time.
11. After 40 minutes, sketch what the tray of ice and sand looks like.

© Earth and Space Science • Weathering and Erosion

Lab 2.1.6A Glacial Erosion

ANALYZE AND CONCLUDE:

1. Describe the environment your tray is in. For example, is the tray in direct sunlight or in shade? What is the temperature of the classroom? Is there any air movement near the tray? Are there any heat sources near the tray, such as people, animals, or running machinery? Did the environment change at all over the course of the experiment? _____

_____

_____

_____

_____

_____

_____

_____

2. Describe the path the ice cube made as it traveled down the tray. _____

_____

_____

_____

3. Compare your data to the data obtained by other groups. What might account for variances among the data collected by different groups? _____

_____

_____

_____

_____

_____

_____

_____

Name: _____ Date: _____

Weathering Observations

WS 2.1.1A

Record the mechanical weathering, chemical weathering, and erosion that you observe.

	Example	Type	What problems is this causing?	How could this be prevented?
1.				
2.				
3.				
4.				
5.				
6.				
7.				
8.				
9.				
10.				

© Earth and Space Science • Weathering and Erosion

Weathering Observations

WS 2.1.1A

Record the mechanical weathering, chemical weathering, and erosion that you observe.

Example	Type	What problems is this causing?	How could this be prevented?
11.			
12.			
13.			
14.			
15.			
16.			
17.			
18.			
19.			
20.			

© Earth and Space Science • Weathering and Erosion

Name: _____ Date: _____

Sculpture Preservation

WS 2.1.2A

Suppose you are a city council member in a town whose economy depends on tourism. You have a number of outdoor sculptures in your town that are important tourist attractions. Each sculpture is made from a different material, including bronze, concrete, granite, and marble. You want to preserve the sculptures for future generations of townspeople and tourists.

Use the information provided and the chart below to analyze the types and effects of weathering on the different materials in the sculptures.
- Rainfall: approximately 36 cm per year
- Snowfall: 10–82 cm per year
- Mean temperature: 19°C
- Temperature range: –21°C to 35°C
- Topography: moderately flat
- Other factors: Dust storms can occur during the spring, summer, and fall.

Material	Type of Mechanical Weathering	Effect of Mechanical Weathering	Type of Chemical Weathering	Effect of Chemical Weathering
Bronze				
Concrete				
Granite				
Marble				

© Earth and Space Science • Weathering and Erosion

Sculpture Preservation

WS 2.1.2A

Complete the second chart to recommend ways of preserving each type of sculpture from the effects of weathering.

Material	Methods of Preservation
Bronze	
Concrete	
Granite	
Marble	

To answer the following questions, consider how others might respond to your preservation efforts and how their opinions and needs might affect how you proceed.

1. Who will value your efforts? Why? _____
_____
_____

2. Who might oppose? Why? _____
_____
_____

3. How can your preservation efforts best satisfy the people in your town? _____
_____
_____
_____

© Earth and Space Science • Weathering and Erosion

Name: _____ Date: _____

The Art of Wind and Water　　　　　　　　　　　**WS 2.1.4A**

Describe at least five features in the area that were formed with the help of wind or water.

1. _____
   _____
   _____
   _____

2. _____
   _____
   _____
   _____

3. _____
   _____
   _____
   _____

4. _____
   _____
   _____
   _____

5. _____
   _____
   _____
   _____

6. _____
   _____
   _____
   _____

7. _____
   _____
   _____
   _____

© *Earth and Space Science* • Weathering and Erosion

Name: _____   Date: _____

Scrape, Rattle, and Roll   WS 2.1.6A

You are going on a trip to see a magnificent natural wonder that God created over time through weathering and erosion processes. To discover your destination, decide if each statement is true or false and circle the answer. Then follow the directions.

		True	False
1.	Mechanical weathering breaks down rocks by wind, water, or acid rain.	forward 10	forward 8
2.	Ice wedging happens in areas that are always below freezing.	back 2	back 4
3.	Chemical weathering happens faster in hot, humid weather.	forward 1	forward 3
4.	Water can be an agent in chemical weathering.	back 3	back 5
5.	A city is more likely to have an acid rain problem than a remote, rural area.	forward 4	forward 3
6.	Granite is often eroded by water to form caves, stalactites, and stalagmites.	back 2	back 1
7.	Iron combines with oxygen to form rust in a process called *oxidation*.	forward 5	forward 2
8.	Mass wasting is much more likely to occur in mountains than on plains because of the topography of these regions.	back 3	back 6
9.	Weathered material moves; eroded material stays in one place.	forward 6	forward 7
10.	A large mass of rock and soil that is sliding rapidly down a slope is called *a landslide*.	back 8	back 7
11.	A large mass of mud that is sliding rapidly down a slope is called *a mud slump*.	forward 4	forward 5
12.	A rock that has been worn away by sand particles has experienced abrasion.	back 4	back 2
13.	A dune is a mound of sand that water deposits onto a beach.	forward 1	forward 7
14.	Glaciers stay in one place.	back 5	back 2
15.	Water that flows over the land is called *groundwater*.	forward 2	forward 10
16.	A narrow ditch eroded by runoff is called *a gully*.	back 6	back 1
17.	Glaciers may create long, tear-shaped mounds called *drumlins*.	forward 4	forward 3
18.	A triangular deposit of sediments where the mouth of a river meets an ocean or lake is called *a delta*.	back 1	back 9

© Earth and Space Science • Weathering and Erosion

Scrape, Rattle, and Roll continued

WS 2.1.6A

START

1. Devil's Marbles, Australia
2. Los Glaciares National Park, Argentina
3. Simien Mountains National Park, Ethiopia
4. White Cliffs of Dover, United Kingdom
5. Kaziranga National Park, India
6. Sahara Desert, Africa
7. Grand Canyon, United States
8. Caves of Aggtelek Karst and Slovak Karst, Hungary
9. Puerto-Princesa Subterranean River National Park, Philippines
10. Darien National Park, Panama
11. Danube Delta, Romania
12. Tajik National Park, Tajikistan
13. Tibetan Plateau, China
14. Strait of Magellan, Chile
15. Victoria Falls, Zambia and Zimbabwe
16. Sognefjord Fjord, Norway
17. Dolomites in the Italian Alps, Italy
18. Bay of Fundy, Canada
19. Mount Kenya National Park, Kenya
20. Lena Pillars Nature Park, Russia
21. Nile River Delta, Egypt
22. Lut Desert, Iran

© Earth and Space Science • Weathering and Erosion

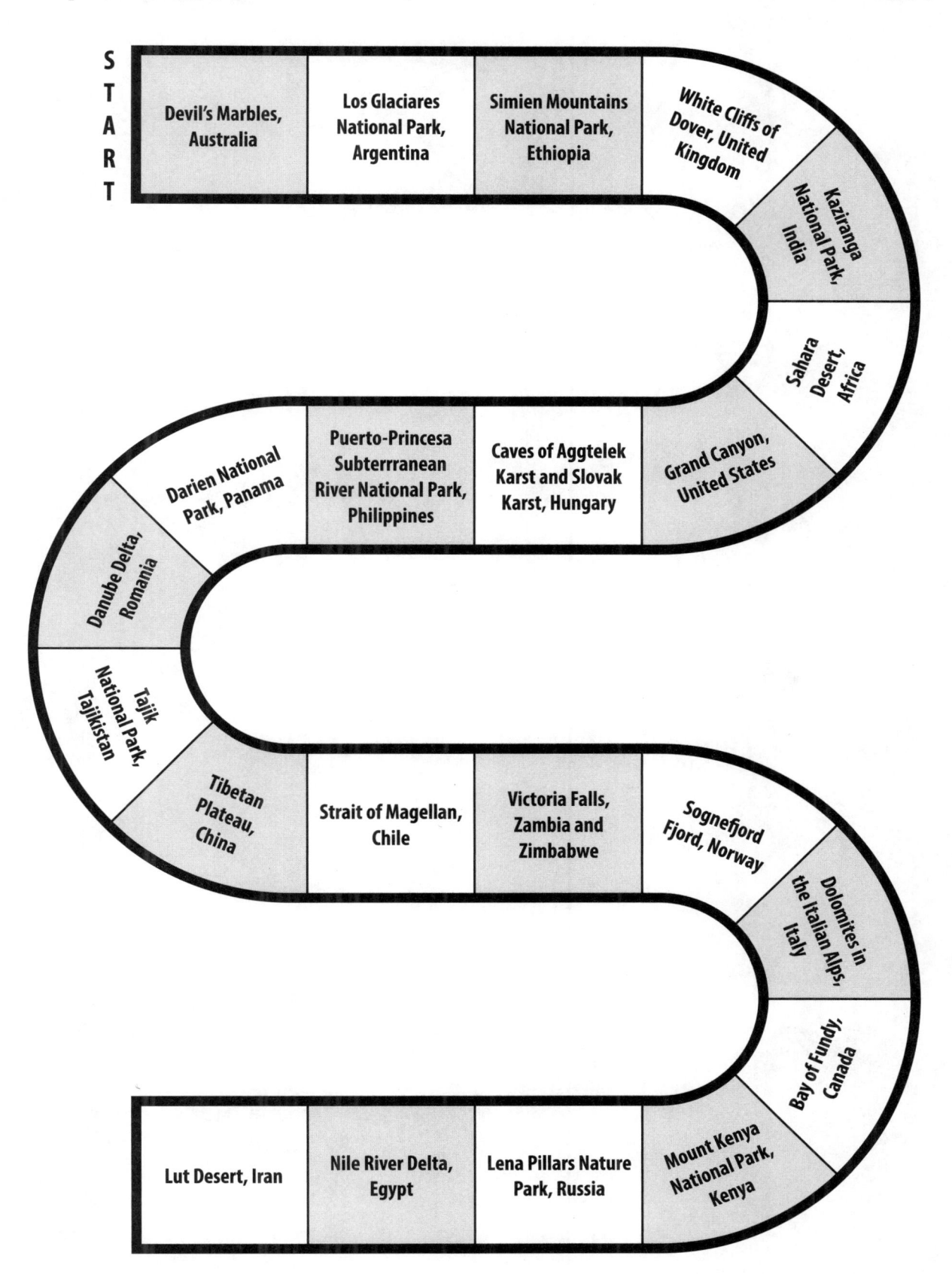

Name: _____   Date: _____

Lab 2.2.1A Pore Space

QUESTION: What determines the amount of pore space in a soil sample?

HYPOTHESIS: _____
_____

EXPERIMENT:

You will need:	• 50 g dry topsoil
• two 250 mL beakers	• spoon

Steps:
1. Pour 150 mL of water into one of the beakers.
2. Place 50 grams of soil into a second beaker.
3. Carefully pour the water into the beaker of soil.
4. Stir gently until bubbles cease to rise to the surface of the water.
5. Measure and record the volume of the water and soil mixture.

- Volume of water: _____
- Volume of dry soil: _____
- Calculated volume of mixture: _____
- Actual volume of mixture: _____
- Difference between calculated and actual volume of mixture: _____

6. Calculate the percentage of pore space filled with air in the soil sample.

% of pore space = (difference in Step 5 ÷ actual volume of mixture) × 100 = _____

ANALYZE AND CONCLUDE:

1. Why do some gardeners think soil that contains a large amount of pore space is better for plant growth than soil with a small amount of pore space? _____
_____
_____
_____

2. Design a controlled experiment to determine the amount of water in soil samples collected around the school.

© Earth and Space Science • Soil

Name: _____ Date: _____

Lab 2.2.1B Capillarity Investigation

QUESTION: How does the capillarity in peat pots relate to capillary action of soil?

HYPOTHESIS: _____
_____

EXPERIMENT:

You will need:	• small dish	• metric ruler
• graduated cylinder	• small peat pot	• stopwatch or watch with second hand

Steps:
1. Measure 75 mL of water and pour it into a shallow dish or pan.
2. Place the peat pot upside down in the water.
3. As the water begins to be absorbed by the peat, measure and record the height of the absorbed water at 1-minute intervals in the data table below.

Time (minutes)	Height of Water (cm)
0	0.0 cm
1	
2	
3	
4	
5	
6	
7	
8	
9	
10	

4. After 10 minutes, remove the pot from the dish or pan and measure how much water is left. Calculate the amount of water that was absorbed by the peat pot.

- Amount of water in the dish at 0 minutes: _____
- Amount of water in the dish at 10 minutes: _____
- Total amount of water absorbed by the peat pot: _____

ANALYZE AND CONCLUDE:

1. Calculate the average rate (cm/min) of capillary action for 10 minutes.

(Rate = distance ÷ time)

- _____ (Rate) = _____ (distance) ÷ _____ (time)

Lab 2.2.1B Capillarity Investigation

2. Draw a line graph illustrating time (*x*-axis) and height of water (*y*-axis).

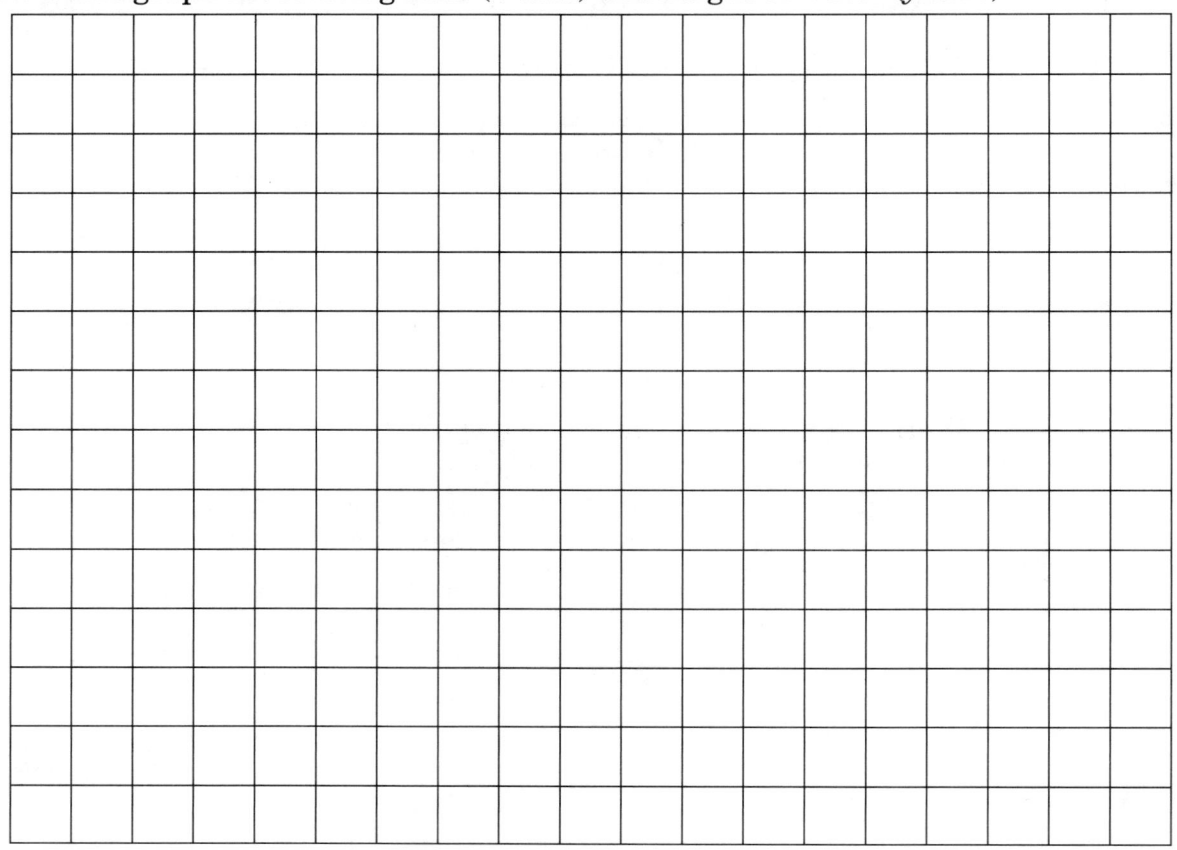

3. Compare how time relates to the height of water as illustrated in the table. _____
_____
_____

4. What type of soil is most similar to peat pots? What type of soil is the most different from peat pots? _____
_____
_____

5. How would knowing about the absorbent abilities of peat be helpful to a farmer or a gardener? _____
_____
_____
_____

© Earth and Space Science • Soil

Name: _____ Date: _____

Lab 2.2.3A Soil Fertility

QUESTION: Will rich topsoil, sandy soil, and clay soil each yield the same fertility result?

HYPOTHESIS: _____
_____

EXPERIMENT:

You will need:	• sandy soil sample	• metric ruler
• 3 planting trays or flowerpots	• clay soil sample	
• rich topsoil sample	• 12 bean or alfalfa seeds	

Steps:
1. Put each type of soil into its own tray or flowerpot and label the containers.
2. Plant 4 seeds in each tray or flowerpot.
3. Place the trays or pots next to each other on a windowsill or in an area that receives plenty of sunlight. Water each soil sample every other day with the same amount of water. Use enough water to moisten the soil. Rotate the trays or pots daily.
4. Observe the trays for two weeks. Record the date that each seed sprouts.

ANALYZE AND CONCLUDE:

1. After two weeks, describe the characteristics of each type of soil, including color, particle size, moisture, and compactness. Record the number of seeds that sprouted in each soil sample.

	Color	**Particle Size**	**Moisture**	**Compactness**	**Number of Seeds**
Topsoil					
Sandy Soil					
Clay Soil					

2. Measure and record the height of each sprout along with the appearance of each sprout.

	Topsoil	**Sandy Soil**	**Clay Soil**
Day 1			
Day 2			
Day 3			

Earth and Space Science • Soil

Lab 2.2.3A Soil Fertility

	Topsoil	Sandy Soil	Clay Soil
Day 4			
Day 5			
Day 6			
Day 7			
Day 8			
Day 9			
Day 10			
Day 11			
Day 12			
Day 13			
Day 14			

3. Which soil was the most fertile? _____

4. What characteristics of this type of soil made it more fertile? _____
_____
_____

5. How would knowing about soil fertility affect a farmer's decisions on what crops to plant? _____
_____
_____
_____

© Earth and Space Science • Soil

Name: _____ Date: _____

Lab 2.2.3B Water-Holding Capacity

QUESTION: Will soil, sand, or silt hold more water?

HYPOTHESIS: _____
_____

EXPERIMENT:

You will need:	• wax marking pencil	• dry silt sample
• 3 funnels	• filter paper	• scale
• 3 ring stands	• potting soil sample	• 100 mL graduated cylinder
• three 250 mL beakers	• dry sand sample	• stopwatch

Steps:
1. Place the funnels in the ring stands. Place a beaker under each funnel. Use the marking pencil to label each beaker as *Soil*, *Sand*, and *Silt*.
2. Form 3 cones from filter paper. Insert them into each of the funnels. Moisten the top of the filter paper so it will adhere to the funnel.
3. Use the scale to measure out 30 g of soil, sand, and silt. Keep the materials separate. Add the materials to each appropriately labeled funnel.
4. Slowly add 100 mL of water to each funnel. Do not allow the water to rise above the top of the filter paper. Record the time the water is added to each funnel. After 10 minutes has elapsed, remove the beaker. Replace the beaker with another receptacle if water is still dripping from the funnel.
5. Measure the quantity of water collected in each beaker. Record your measurements and observations below.

	Initial Time	**Final Time**	**Amount of Water Collected**
Soil			
Sand			
Silt			

ANALYZE AND CONCLUDE:

1. Which soil held the most water? Why? _____
_____
_____

2. Would a combination of the soils be the best environment for plants? Why?
_____
_____

Earth and Space Science • Soil

Name: _____ Date: _____

The Layered Look   WS 2.2.2A

1. Dig a hole about 1 m deep in three different regions: a wetland, a wooded area, and a grassy area.
2. Measure and record the pH and temperature of each layer right after digging.
3. Mark where the soil changes color and indicate what each layer looks like in the tables below. Measure and record the depth of each layer. (Note: All six layers may not be observed.)

Area 1 _____

	Appearance	Color	Texture	Depth (cm)	pH	Temperature (C°)
O Horizon						
A Horizon						
E Horizon						
B Horizon						
C Horizon						
R Horizon						

What is the parent material of this soil sample?

Area 2 _____

	Appearance	Color	Texture	Depth (cm)	pH	Temperature (C°)
O Horizon						
A Horizon						
E Horizon						
B Horizon						
C Horizon						
R Horizon						

What is the parent material of this soil sample?

© Earth and Space Science • Soil

The Layered Look

WS 2.2.2A

Area 3						
	Appearance	**Color**	**Texture**	**Depth (cm)**	**pH**	**Temperature (C°)**
O Horizon						
A Horizon						
E Horizon						
B Horizon						
C Horizon						
R Horizon						

What is the parent material of this soil sample?

© Earth and Space Science • Soil

Name: _____   Date: _____

Law of Superposition
WS 2.2.4A

Use the illustration below to answer the questions.

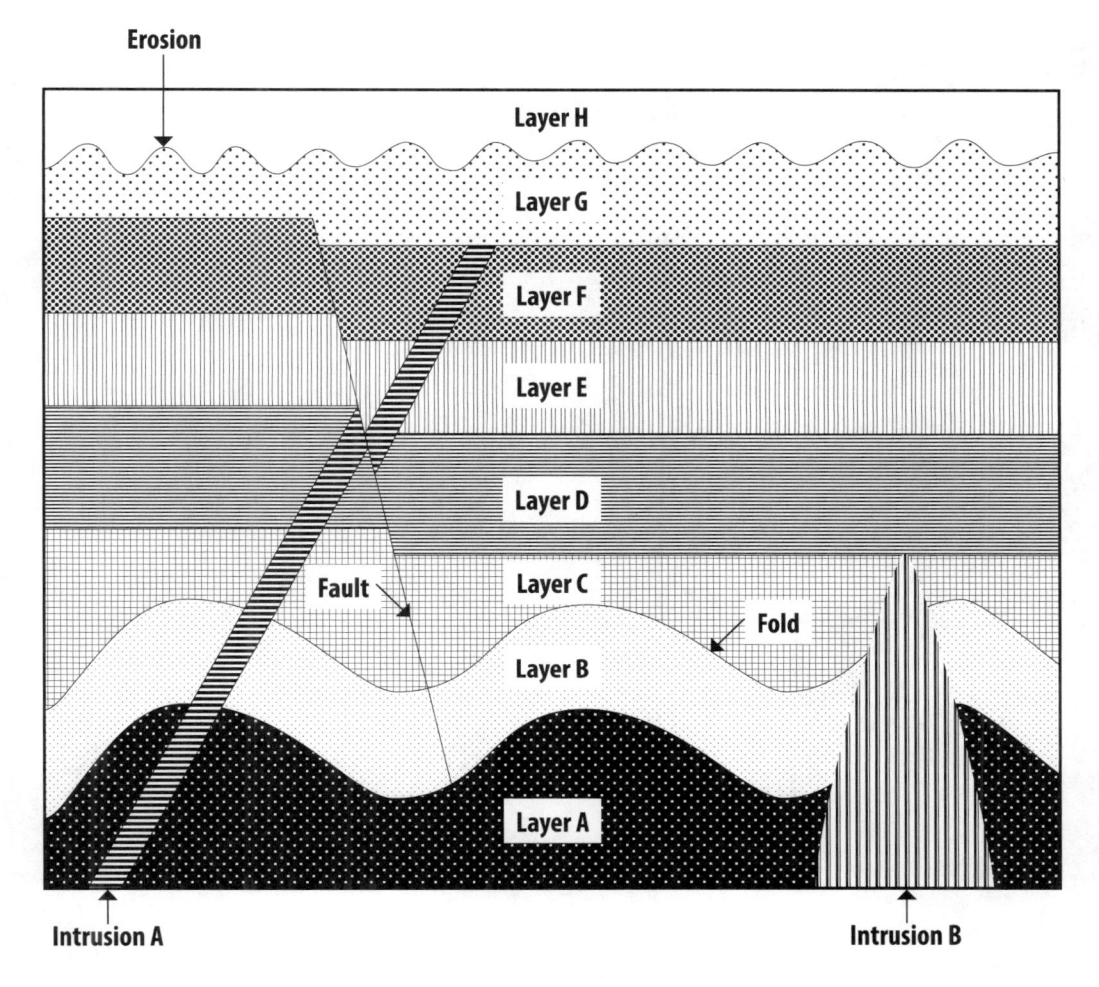

Erosion → Layer H
Layer G
Layer F
Layer E
Layer D
Fault → Layer C
Layer B ← **Fold**
Layer A

↑ **Intrusion A**   ↑ **Intrusion B**

1. Is Intrusion A (igneous rock) younger or older than Layers C and D (sedimentary rock)? _____

2. Did the fault occur before or after Layers A–F were laid down? _____

3. Was Layer G laid down before or after the fault occurred? _____

4. Did the fault occur before or after Intrusion A? _____

5. Did the folding occur before or after Layer C was laid down? _____

6. Did the erosion occur before or after Layer H was laid down? _____

7. Did Intrusion B occur before or after the folding? _____

8. Which is older—Intrusion A or Intrusion B? _____

_____

_____

© Earth and Space Science • Soil

Name: _____ Date: _____

Lab 3.1.1A Isostasy

QUESTION: What would cause different parts of the crust to float at different depths in the asthenosphere?

HYPOTHESIS: _____

_____

EXPERIMENT:

You will need:	• waterproof clear plastic tub	• ice disk
• 2 plastic petri dishes	• cold water	
• stackable metric weights	• metric ruler	

Steps:
1. Fill the tub about half full of water. The water represents the asthenosphere.
2. Float the empty petri dishes on the water. The dishes represent the continental crust. Make a drawing below to show where the bottom of the dishes are in relation to the surface of the water.
3. Stack the weights into two columns. One stack should be about twice as high as the other. These will need to sit on the floating petri dishes, so do not make them too tall. Record the height and mass of each stack. _____

_____

4. Carefully place a stack of weights in the center of each petri dish. The taller stack represents a mountain and the shorter stack represents a flatland.
5. Look through the side of the tub and draw what you see now, showing where the bottom of each dish is in relation to the surface of the water.
6. Ask your teacher for an ice disk. The disk represents the oceanic crust. Place the disk in the water. Make a drawing to show where the bottom of the disk is in relation to the surface of the water.

Floating Crust

Empty Petri Dishes	**Stacked Petri Dishes**	**Ice Disk**

© Earth and Space Science • Crust Movement

Lab 3.1.1A Isostasy

ANALYZE AND CONCLUDE:

1. What does the empty petri dish represent? _____

2. What does the ice disk represent? _____

3. Which one sits lower in the water? _____

4. How is the ice disk like the oceanic crust? _____

_____

_____

5. Which dish sat lower in the water? Why? _____

_____

_____

6. Explain how this lab demonstrates the theory of isostasy. _____

_____

_____

_____

_____

_____

_____

Name: _____ Date: _____

Lab 3.1.3A The Effects of Stress

QUESTION: How do different types of stress affect rock formations?

HYPOTHESIS: _____

_____

EXPERIMENT:

You will need:	• 4 colors of clay	• knife

Steps:
1. Shape each color of clay into a long flat rectangle about 3 cm thick. Each color should be the same shape and size.
2. Layer the clay rectangles.
3. Cut the block into four equal sections.
4. Take one block and push in from opposite sides. Record the changes made to Block 1. _____

5. Take the second block and slowly pull the block from opposite sides. Record the changes made to Block 2. _____

_____

6. Take the last two blocks and rub one side of one block against a side of the other block. Record the changes made to Blocks 3 and 4. _____

_____

7. Cut through each block vertically. Sketch the layers in each block.

Block 1	Block 2	Block 3	Block 4

ANALYZE AND CONCLUDE:

1. What type of stress was applied in Step 4? _____
2. Record the type of stress applied in Step 5. _____
3. What type of stress was applied in Step 6? _____

© Earth and Space Science • Crust Movement

Lab 3.1.3A The Effects of Stress

4. Describe how each type of stress deforms the block. Include the terms *volume*, *density*, and *shape*. _____
_____
_____
_____
_____
_____

5. What properties of the clay affected its deformation? _____
_____
_____
_____

6. What factors could change how the clay deformed? _____
_____
_____
_____
_____
_____

7. How does what you observed in the clay deformation relate to what happens when rock is under stress? _____
_____
_____

Name: _____ Date: _____

Graph the Continents   WS 3.1.1A

Create a pie graph showing the percentage of land area that each continent occupies.
1. Find the area of each continent.
2. Add all the areas together to get total land area on the planet.
3. Figure the percentage of the area that each continent occupies.
4. Make a pie graph to show the percentage that each continent occupies. Choose a different color for each continent.

Area of Each Continent

Continent	Area in km^2	Percentage of Earth's Landmass

1. Which continent has the greatest percentage of the earth's land? _____

2. Which continent has the smallest percentage of the earth's land? _____

3. What type of rock makes up most of the continental crust? _____

4. Is it more dense or less dense than oceanic crust? _____

© Earth and Space Science • Crust Movement

Name: _____ Date: _____

It's a Puzzle

WS 3.1.1B

On a piece of white paper, trace slightly outside the edge of each continent as shown on the map. Cut out each shape you traced. Arrange the continents on a large sheet of construction paper to fit together into one supercontinent. Try several different arrangements. Glue the cutouts onto the construction paper in the shape you think is best.

1. Were you able to make a supercontinent? _____

2. If not, what do you think caused the problem? _____

_____

3. Do you believe all the continents started as one supercontinent? Why? _____

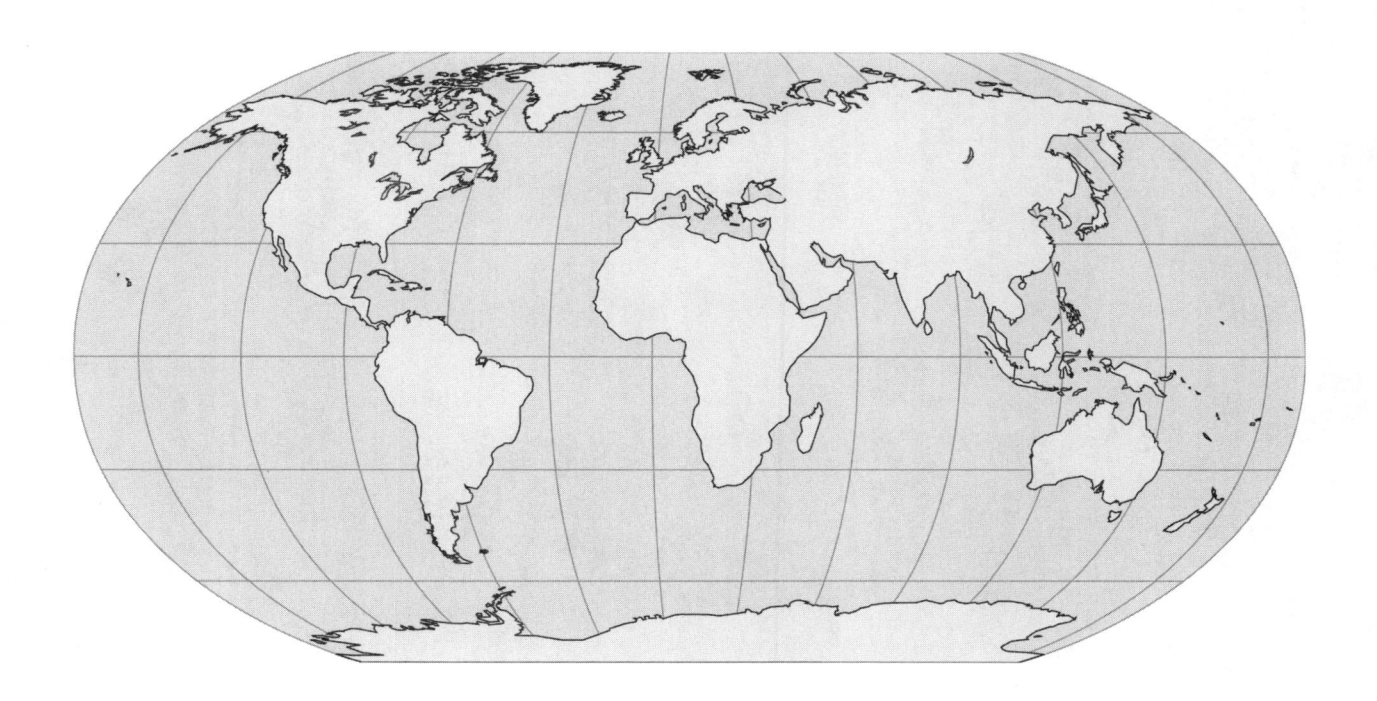

_____
_____
_____
_____
_____
_____
_____

© Earth and Space Science • Crust Movement

Name: _____   Date: _____

Digging Deeper   WS 3.1.1C

Fill in the chart below with information gathered from previous class research.
1. Carefully choose which information you believe is fact and which information is a conclusion. You may have more than one conclusion for one fact.
2. Work with students in other groups to get all the information needed for the chart.

Scientific Evidence

Facts	Conclusions

1. Which column gives absolute truth? _____

2. Which column shows opinions based on worldviews? _____

3. Do scientists ever come up with different conclusions when given the same facts? Give an example. _____
_____

4. How should disagreements be handled? _____
_____

5. What direct evidence would prove that the continents were originally one giant supercontinent? _____
_____
_____

6. Does this kind of evidence exist? _____

7. Is it possible to know with absolute certainty that all continents started as one?

© Earth and Space Science • Crust Movement

Name: _____ Date: _____

Under Stress   WS 3.1.3A

List the materials provided. Record what happens to each item as you apply the different types of stress.

Stress

Item	Compressional	Tensional	Shearing

1. What materials deformed under compressional stress? Were these items ductile, elastic, or brittle? _____
_____
_____

2. What materials deformed under tensional stress? Were these items ductile, elastic, or brittle? _____
_____
_____

3. What materials deformed under shearing stress? Were these items ductile, elastic, or brittle? _____
_____

4. Describe the characteristics of the materials that deformed the most. _____
_____

5. Describe the characteristics of the materials that deformed the least. _____
_____

© Earth and Space Science • Crust Movement

Name: _____ Date: _____

Mountain Range Scavenger Hunt

WS 3.1.4A

Research the following mountain ranges and fill in the information on the chart. For the last row, research a mountain range of your choice.

Mountain Range	Countries	Length (km)	Highest Peak (m)	Name of Climber and Date of First Ascent	Latitude and Longitude of Highest Peak
Alps					
Andes Mountains					
Cascade Range					
Great Dividing Range					
Himalayas					
Rocky Mountains					
Rwenzori Mountains					

1. Which range is the longest? _____

2. Which mountain is the highest? _____

3. What mountains have you been to? _____

4. Which mountain range is closest to your home? _____

5. Name one range that is made of folded mountains. _____

6. Name one range that is made of fault-block mountains. _____

© Earth and Space Science • Crust Movement

Name: _____   Date: _____

Mountain Perspectives   WS 3.1.4B

What would the following people see if they were all looking at the same mountain?

1. an artist _____
_____

2. a theologian _____
_____

3. a rock climber _____
_____

4. a biologist _____
_____

5. a geologist _____
_____

6. a road builder _____
_____

Answer the following questions on a separate piece of paper:

7. What questions would each of these people ask about the mountain?

8. What tools would each use to see the mountain more fully?

9. Is the theologian the only one who might be seeing with the eyes of faith?

10. Does any one way of seeing offer the whole truth about the mountain?

11. What can each perspective see that the others might miss?

12. What would be lost if you adopted only one of these ways of seeing and got rid of the others?

13. Which of these ways of seeing can be expressions of care for, or appreciation of, the mountain?

14. How can learning science deepen your care for the world?

15. How can humility allow people to appreciate the value of different perspectives?

Name: _____ Date: _____

Lab 3.2.2A Earthquake Energy

QUESTION: What happens to energy as it is spent over a distance?

HYPOTHESIS: _____
_____

EXPERIMENT:

You will need:	• chalk or masking tape	• rubber band
• desk	• disposable cup	• tape measure
• meterstick	• 50 g sand	

Steps:
1. Measure a 2 m distance from the edge of your team's desk. Mark the distance with chalk or masking tape.
2. Fill a large disposable cup with 50 g of sand.
3. Set the cup on the edge of the desk closest to the chalk line.
4. One team partner should place a rubber band over his finger and stretch it back. The other partner should measure and record the length of the stretched rubber band and make sure that the first partner's fingertip is aligned with the edge of the desk and facing the cup.
5. Keeping the rubber band in place, aim at the cup and release the rubber band. Measure and record the distance the cup moved back from the edge of the table. In the same space, note whether the cup fell over.
6. Replace the cup and any sand that spilled from it.
7. Increase the distance from the desk by 1 m.
8. Repeat Steps 1–6, making sure to stretch the rubber band the same length as in the first trial.
9. Repeat the experiment two more times, each time increasing the distance from the desk by 1 m and always stretching the rubber band to the same length as in the first trial.

Trial	Length of Stretched Rubber Band	Distance from Desk	Distance Cup Moved
1.		2 m	
2.		3 m	
3.		4 m	
4.		5 m	

© Earth and Space Science • Earthquakes

Lab 3.2.2A Earthquake Energy

ANALYZE AND CONCLUDE:

1. Create a line graph of the results. On the *x*-axis plot the distance from the desk, and on the *y*-axis plot the distance the cup moved.

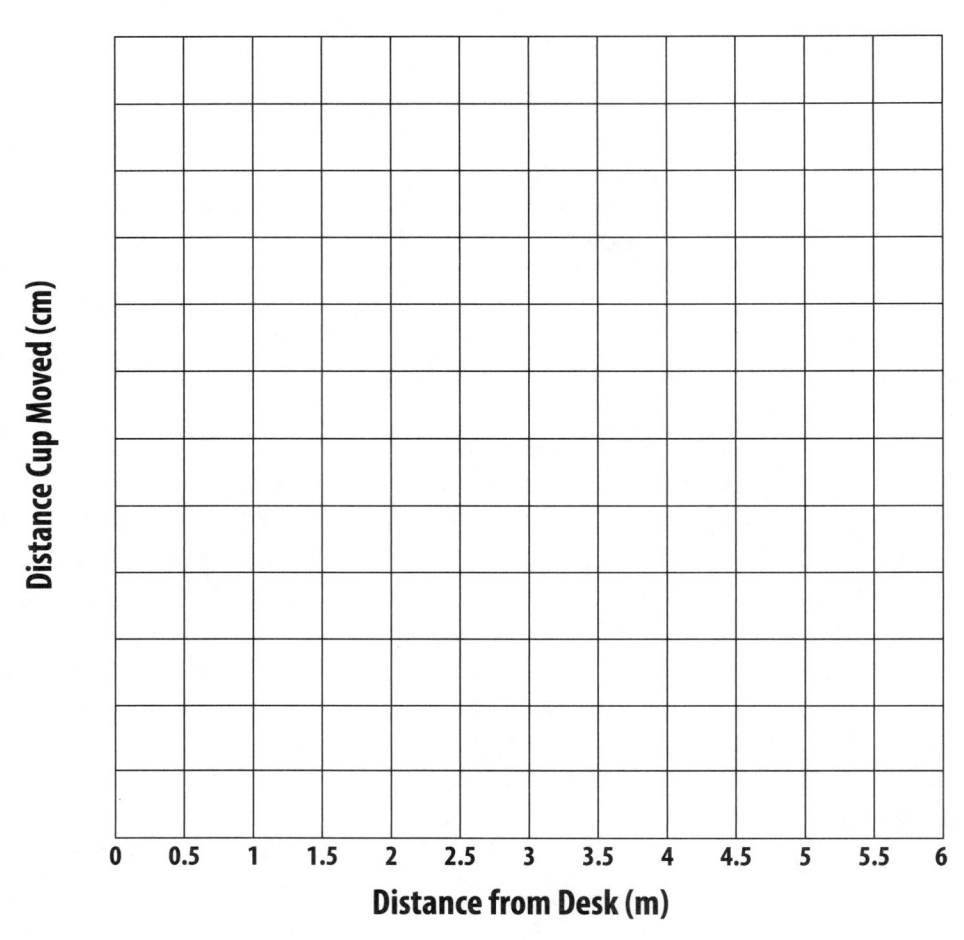

Distance Cup Moved (cm)

0   0.5   1   1.5   2   2.5   3   3.5   4   4.5   5   5.5   6

Distance from Desk (m)

2. What is the relationship between the distance the rubber band traveled and the distance the cup moved? _____

_____

3. What accounts for the relationship between the distance the rubber band traveled and the distance the cup moved? _____

_____

_____

4. How does this experiment correspond to an earthquake's depth and the amount of its force that is felt on the earth's surface? _____

_____

_____

© *Earth and Space Science* • Earthquakes

Name: _____ Date: _____

Lab 3.2.4A Seismograph

QUESTION: How does a seismograph work?

HYPOTHESIS: _____

_____

EXPERIMENT:

You will need:	• ring stand	• tape
• string	• meterstick	• felt-tip pen
• lead weight	• graph paper	

Steps:
1. Find a partner and use string to suspend a lead weight from a ring stand so the weight is approximately 5 cm from the work surface.
2. Place a piece of graph paper underneath the ring stand.
3. Tape the pen to the weight so the point of the pen just touches the paper.
4. Have one partner jump up and down while the other partner slowly and steadily pulls the paper forward.
5. Label the graph paper with the name of the person who jumped. Measure the distance the person was from the seismograph with a meterstick.
6. Replace the graph paper with a fresh piece.
7. Trade jobs with your partner and repeat Steps 4–5.
8. Try jumping in different locations. For each location, label the graph paper with the type of force exerted and the distance the force was from the seismograph.

ANALYZE AND CONCLUDE:

1. Why must the paper be moving and why must it move at a steady rate? _____

_____

_____

2. What did you observe regarding the marks on the graph and the approximate force of the jumper? _____

3. What did you observe regarding the marks on the graph and the distance between the jumper and the seismograph? _____

_____

4. Why is more than one seismograph needed to accurately describe an earthquake?

_____

_____

_____

© Earth and Space Science • Earthquakes

Name: _____ Date: _____

Earthquake Zone Map

WS 3.2.2A

Use the map to plot recent earthquakes. Use a blue dot for shallow earthquakes (0–70 km deep), a yellow dot for intermediate earthquakes (70–300 km deep), and a red dot for deep earthquakes (more than 300 km deep). List the earthquakes' locations on the next page and write each earthquake's number next to its dot on the map. Draw and label the tectonic plate boundaries. Draw arrows to show the direction each plate is moving. Answer the questions on the next page.

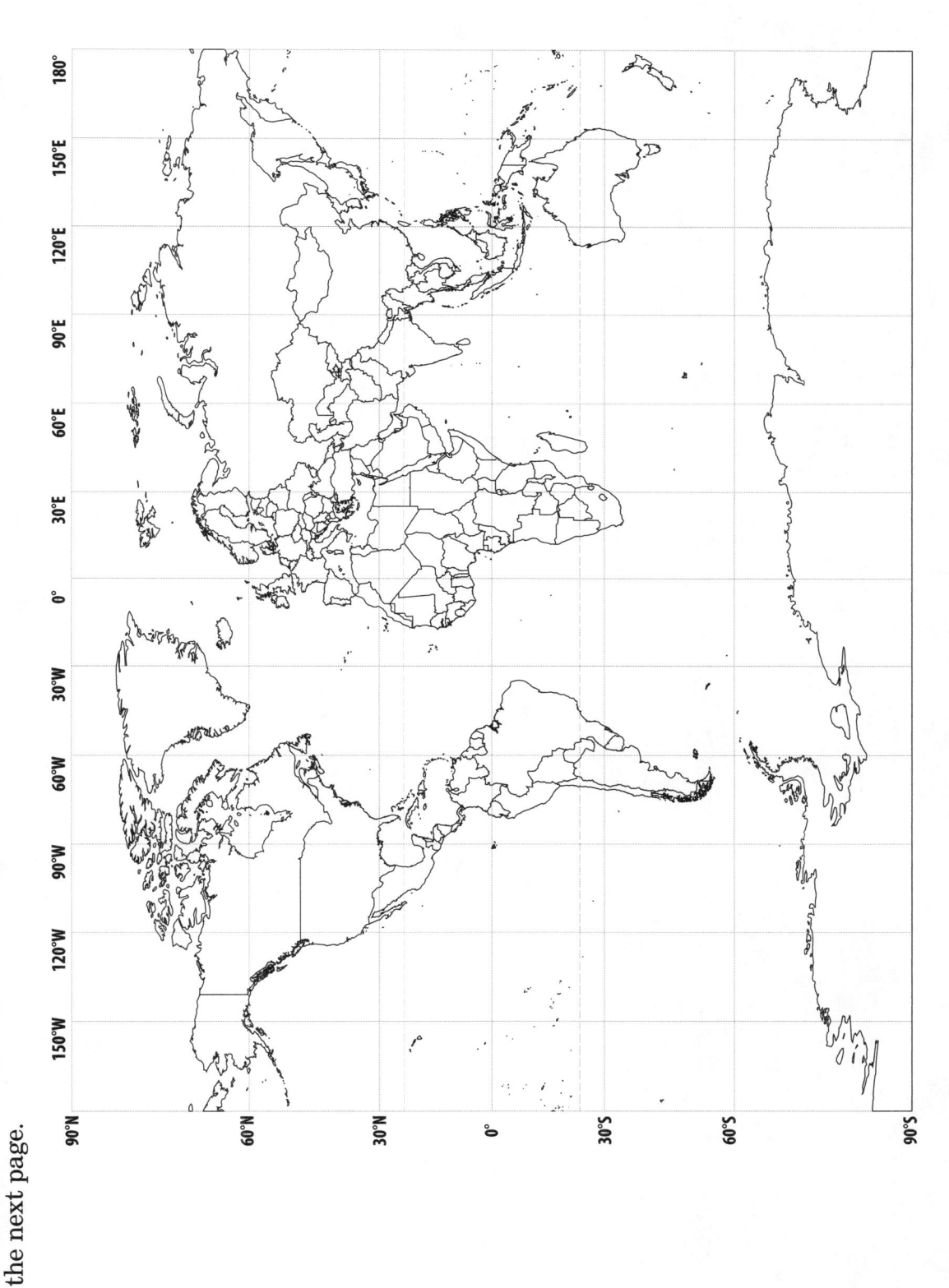

© *Earth and Space Science* • Earthquakes

Name: _____ Date: _____

Earthquake Zones

WS 3.2.2B

Earthquakes

	Location			Location
1.			11.	
2.			12.	
3.			13.	
4.			14.	
5.			15.	
6.			16.	
7.			17.	
8.			18.	
9.			19.	
10.			20.	

1. Along which type of boundary do most of the shallow earthquakes lie? _____
_____

2. Along which type of boundary do most of the intermediate earthquakes lie?
_____

3. Along which type of boundary do most of the deep earthquakes lie? _____
_____

4. What relationship do you see between plate boundaries and earthquake locations?
_____
_____

© Earth and Space Science • Earthquakes

Name: _____ Date: _____

Analyze a Strike-Slip Fault   WS 3.2.3A

The San Andreas Fault is the main boundary between the Pacific Plate and the North American Plate. It is one of the major strike-slip boundaries in the world. The two plates are moving, but not at the same rate. The difference in rates is about 6 cm per year. Use this information to help you answer the questions listed below.

1. Under what conditions might destructive earthquakes result? _____
_____
_____
_____

2. How many years and months will it take a plate to move 1 m at the current rate?
_____

3. If a stream crosses the fault, how might it change after the fault moves? _____
_____
_____
_____

4. In which of the following scenarios is a large earthquake more likely to occur: after a year when the plates move 6 cm or after a year when the plates do not move?

Explain your answer. _____
_____
_____
_____
_____

5. How could you determine how much the plates move in any given year? _____
_____
_____
_____
_____
_____
_____

Name: _____   Date: _____

Measuring Earthquakes Graph   WS 3.2.4A

Use the graph to make a scatter plot showing the intensity and magnitude of at least 20 earthquakes. Intensity measurements should come from the epicenter or as close to it as possible. Use the table to identify a key for each earthquake as well as each earthquake's epicenter and depth below the surface. Use the completed graph and table to answer the questions.

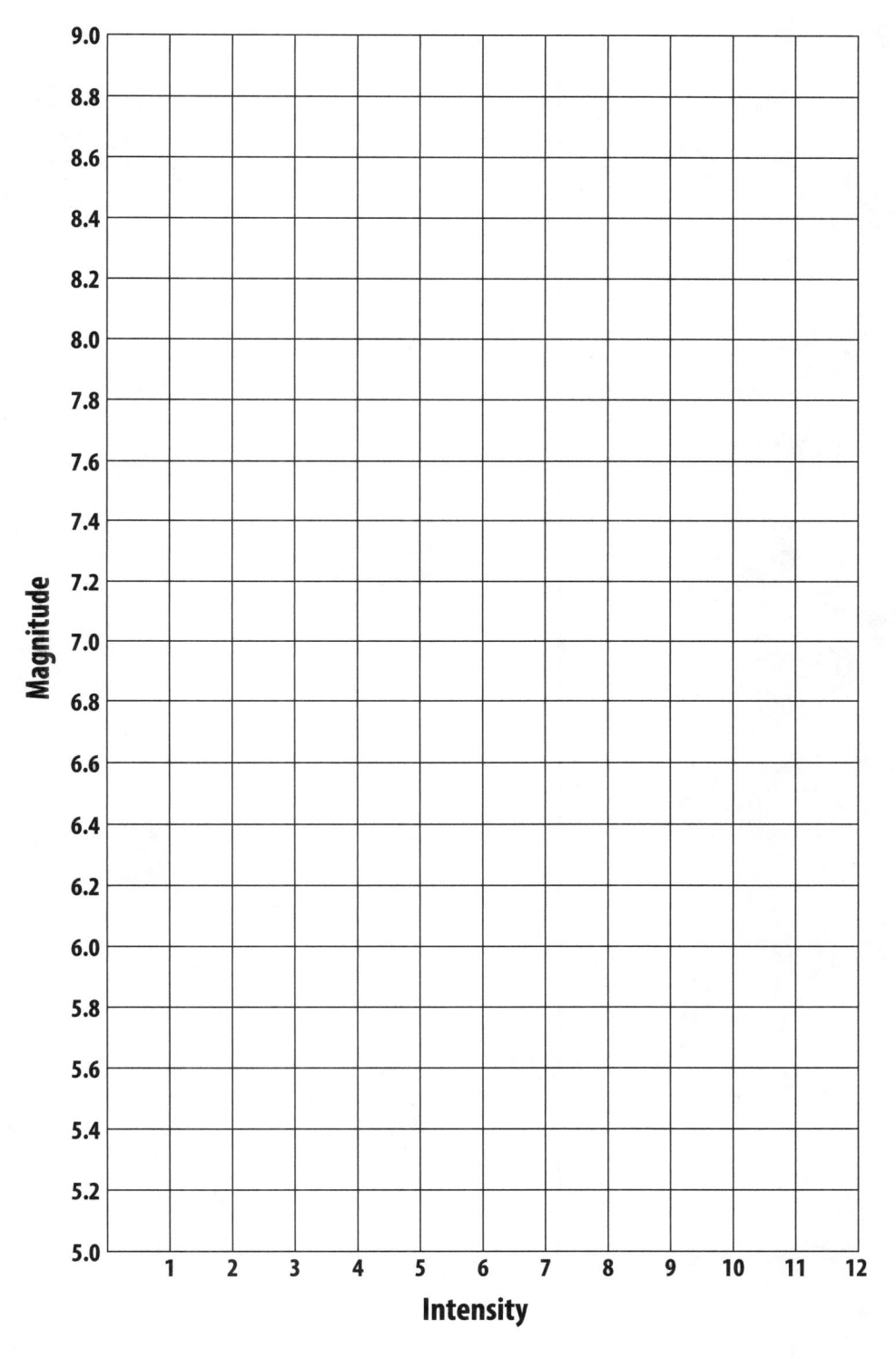

Magnitude (y-axis: 5.0 to 9.0 in intervals of 0.2) vs. **Intensity** (x-axis: 1 to 12)

Name: _____   Date: _____

Measuring Earthquakes WS 3.2.4B

Earthquakes

Key	Epicenter	Depth		Key	Epicenter	Depth

1. What intensity scale did you use? _____

2. What magnitude scale did you use? _____

3. What relationship do you see between intensity and magnitude? _____
_____

4. Are there earthquakes with the same intensity but different magnitudes? _____
What might explain this? _____
_____

5. Based on the graph and table, what factors affect an earthquake's intensity?
_____
_____

© Earth and Space Science • Earthquakes

Name: _____   Date: _____

Earthquake Prediction   WS 3.2.5A

Scientists have not yet discovered an accurate method of predicting earthquakes. Such a method would be a blessing, but it would also present challenges. Consider the following situation and answer the questions, thinking about your responsibility as a Christian.

You have invented a new method of predicting earthquakes. It has not yet been perfected, but it has proven accurate enough with small earthquakes that you have begun to publicize warnings. Six months ago, your instruments warned that a 5.1 earthquake had an 85% chance of striking within six hours, and you warned the public. Lots of people were late for school and work because they took the time to put their breakable objects in safe places, and many stores closed so that they could secure their merchandise and avoid the risk of things falling on their customers. No earthquake came. Three months later, your instruments warned of an 80% chance of a 6.0 earthquake the next day. Many schools were cancelled, stores closed again, and many businesses closed for the day. Again, no earthquake came. People are becoming angry about the false alarms, and many no longer trust your system. Your system worked when you were in the initial testing stages, but now you wonder if it is flawed. Your funding has almost run out, and if you issue any more false alarms, you fear that it will not be renewed. If this happens, you will never find a foolproof system to predict earthquakes—one that could save many lives. Now your instruments are warning of a 75% chance of an 8.0 earthquake tomorrow. What should you do?

Answer the following questions on a separate piece of paper:
1. Do you think that your instruments are flawed or that in each of the previous cases the absence of the earthquake just "beat the odds"?
2. What would be worse—to fail to report the possibility of a devastating earthquake and have one happen or to report it, have much of the public ignore the warning because they do not trust it, cause panic in the rest of the population, and then not have the earthquake occur, which would reduce both your funding and the chances of ever predicting earthquakes?
3. If you decide to warn the public, how do you do so without causing some of the people to panic while making sure that the other people realize that the possibility is real? How do you remain sensitive to the fact that you have been wrong before and have inconvenienced the public?

Name: _____ Date: _____

## Earthquake Prevention	WS 3.2.5B

Scientists have not discovered a method of preventing earthquakes. Such a method would be a blessing, but it would also present challenges. Consider the following situation and answer the questions, thinking about your responsibility as a Christian.

You have invented a method of preventing large earthquakes. Your prediction instruments have told you that the fault lying under your heavily populated city has a 50% chance of undergoing a major earthquake (at least 7.1) within two years. Your prevention method involves blasting the part of the fault where the most pressure is built up. The city's stadium and several houses stand on top of the part of the fault where you think most of the pressure has built up. In order to execute your method, you will have to destroy the stadium. The city just won the campaign to host the world championships there in six months. Everyone wants you to wait until the world championships are over before you start. They want to experience the excitement and the extra business that the championships will bring.

Answer the following questions on a separate piece of paper:
1. Without causing panic, how will you remind the public that if a major earthquake strikes the stadium and the houses, the rest of the city will be heavily damaged?
2. How much time will you need to allow people to move out of their houses and make other arrangements? How will they be compensated?
3. There is a small chance that blasting the fault could trigger the earthquake that you are trying to prevent. Is the risk worth it? Why?
4. Is it worth the risk to wait six months to enact your plan, or should you go ahead as soon as the houses in question are vacated? Should strong public opinion count when lives and property are at stake? Why?
5. If you decide to act before the six months are up, how would you calm the people who are angry?
6. If you destroy the houses and stadium and discover that your calculations were wrong and that part of the fault is not the place where the most pressure is built up, how would you address the issue?

Name: _____ Date: _____

Lab 3.3.1A How Volcanoes Form

QUESTION: What creates the different features of a volcano?

HYPOTHESIS: _____
_____

EXPERIMENT:

You will need:	• plastic tubing	• clamp to fit plastic tubing
• box	• balloon	• flour
• newspaper	• tape	• metric ruler

Steps:
1. Line the box with newspaper. Punch a hole through the middle of the bottom of the box and the newspaper. The hole should be just large enough to allow the plastic tubing through.
2. Pass the tubing through the hole.
3. Tape and seal the deflated balloon on the end of the tubing inside the box.
4. Carefully pile flour on top of the balloon and the surrounding area.
5. Slowly blow through the tubing from the other end to inflate the balloon to approximately 10 cm in diameter. Then clamp the outside end of the tube to hold the air in the balloon.
6. If any part of the balloon is showing through the flour, cover it and mold the flour into the shape of a volcano cone.
7. Illustrate and label your observations.
8. Release the clamp to deflate the balloon.
9. Illustrate and label your observations.

Balloon Volcano Observations

Inflated Balloon Volcano	**Deflated Balloon Volcano**

© Earth and Space Science • Volcanoes

Lab 3.3.1A How Volcanoes Form

ANALYZE AND CONCLUDE:

1. How did the volcano form? _____
_____

2. What part of a volcano does the balloon represent? _____
_____

3. What formed the crater in the volcano? _____
_____

4. How is this similar to a crater forming in a real volcano? _____
_____
_____
_____

5. What factors might affect the size of a volcano's crater? _____
_____
_____

6. Design an experiment to test this hypothesis. _____
_____
_____
_____
_____
_____
_____
_____
_____
_____
_____
_____

© Earth and Space Science • Volcanoes

Name: _____ Date: _____

Lab 3.3.1B How Magma Moves

QUESTION: What determines where and how magma will move in a volcano?

HYPOTHESIS: _____

_____

EXPERIMENT:

You will need:	• 2 hot pads	• large syringe
• small bowl of set, unflavored gelatin	• large bowl	• chocolate syrup
• aluminum pan	• 2–4 bricks or large blocks	

Steps:
1. Punch 3 or 4 holes in the pan, just large enough for the syringe to go through.
2. Heat about 2 cups of water. Using hot pads, pour the hot water into the large bowl. Dip the small bowl in the hot water to loosen the gelatin.
3. When the gelatin has slightly loosened from the bowl, place the pan on the gelatin bowl. Turn the pan and bowl over together to release the gelatin from the bowl. The set gelatin represents a volcano.
4. Place the pan with the gelatin volcano on top of the bricks. This pan should be raised up enough to get the syringe underneath and through the holes.
5. Fill the syringe with chocolate syrup, which represents the magma.
6. Predict what will happen when you inject the syrup into the gelatin volcano. How will it travel through the gelatin volcano? Will it break through the surface? Where? _____

_____

_____

7. Insert the syringe through one of the holes in the pan and into the gelatin volcano. Very slowly inject the syrup.

8. Record your observations. _____

_____

_____

_____

9. Repeat Step 7 using a different hole in the pan.

10. Record your observations. _____

_____

_____

_____

© Earth and Space Science • Volcanoes

Lab 3.3.1B How Magma Moves

ANALYZE AND CONCLUDE:

1. What caused the chocolate syrup to move through the gelatin? _____
_____

2. What causes magma to move through a volcano? _____
_____

3. What do you think determined the path the syrup would take through the gelatin?
_____
_____
_____

4. What factors affect the path magma takes through a volcano? _____
_____
_____
_____
_____
_____
_____
_____

Name: _____   Date: _____

Lab 3.3.3A Viscosity

QUESTION: How does viscosity affect lava flow?

HYPOTHESIS: _____
_____

EXPERIMENT:

You will need:	• vegetable oil	• tape
• corn syrup	• waxed paper	• metric ruler
• molasses	• pipettes	• stopwatch
• isopropyl alcohol	• 20 cm × 30 cm cardboard	

Steps:
1. Cut the waxed paper to fit the cardboard and tape it to the cardboard.
2. Place the cardboard on the table. Using a separate pipette for each liquid, place two drops of each liquid on the waxed paper. The drops should be placed in a straight line parallel with a short edge.

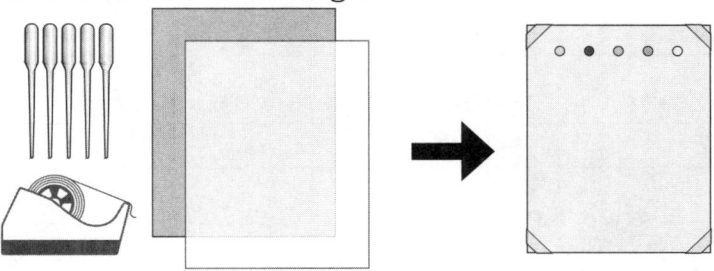

3. Predict which liquid will reach the bottom of the cardboard first and which will reach the bottom of the cardboard last.

_____
_____

4. Lift the end of the cardboard with the liquid drops. Raise it about 45° off the table and watch the fluids.
5. Record your results.

Observations

First (least viscous)	Second	Third	Fourth	Fifth (most viscous)

6. Remove the waxed paper from the cardboard.
7. Tape a new sheet of waxed paper to the cardboard.
8. Place two drops of corn syrup at the top of the cardboard. Predict how long it will take the corn syrup to reach the bottom of the cardboard and record your prediction in the next Observations table.
9. Lift the cardboard 45° off the table and start the stopwatch. Record the time it takes the corn syrup to reach the bottom of the cardboard.

© Earth and Space Science • Volcanoes

Lab 3.3.3A Viscosity

10. Repeat Steps 8–9 for the other four liquids.

Observations

	Corn Syrup	Molasses	Isopropyl Alcohol	Vegetable Oil	Water
Prediction Time					
Actual Time					

ANALYZE AND CONCLUDE:

1. In Step 3, which liquid did you predict to be the least viscous? How does your prediction compare with your results? _____
_____

2. In Step 5, which liquid was the most viscous? _____

3. Was the most viscous liquid in Step 5 the slowest liquid to reach the bottom in Step 10? Why? _____
_____

4. How do your results in the first trial compare with your results in the second trial?
_____

5. Is the same liquid the least viscous in both cases? _____
Is the same liquid the most viscous in both cases? _____

6. If your results are not the same, explain why you think that is. _____
_____
_____

7. How does silicate content affect the viscosity of lava? _____
_____
_____

Name: _____ Date: _____

Decade Volcanoes WS 3.3.1A

In the 1990s, the United Nations initiated the International Decade for Natural Disaster Reduction, a project to reduce natural disasters. Sixteen volcanoes were chosen to be part of the project by using a list of qualifications. Each volcano had more than one volcanic hazard, had experienced recent geological activity, was located in a populated area, was politically and physically accessible for study, and had local support for the work. These volcanoes were labeled *the Decade Volcanoes*.

1. Research each volcano. Find its country and location.
2. Plot each volcano on the world map.

Decade Volcanoes

Volcano	Country	Latitude and Longitude
Avachinsky-Koryaksky		
Colima		
Galeras		
Mauna Loa		
Mount Etna		
Mount Merapi		
Mount Nyiragongo		
Mount Rainier		
Mount Vesuvius		
Mount Unzen		
Sakurajima		
Santa Maria		
Santorini		
Taal Volcano		
Teide		
Ulawun		

© *Earth and Space Science* • Volcanoes

Name: _____ Date: _____

Decade Volcanoes Map

WS 3.3.1B

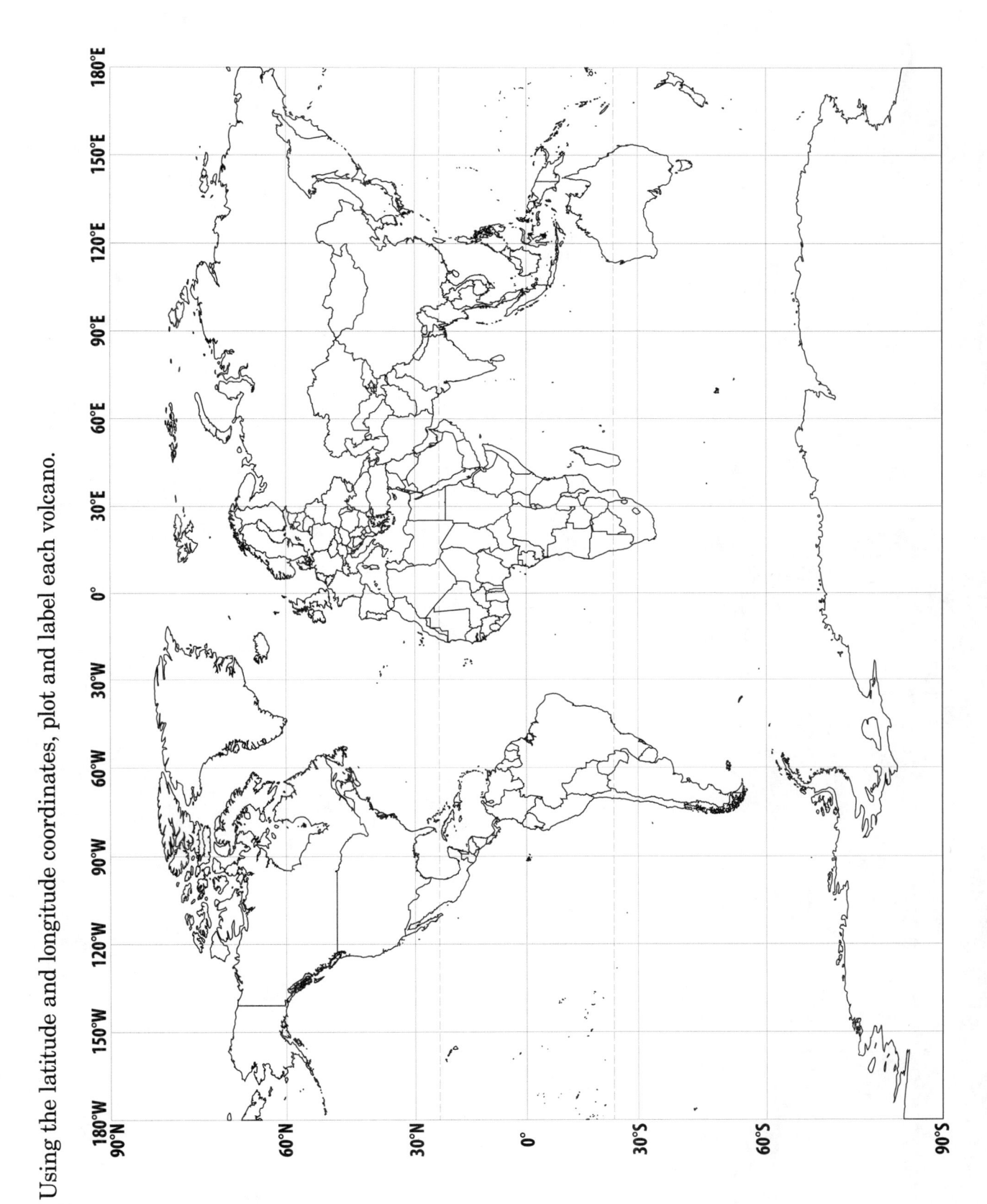

Using the latitude and longitude coordinates, plot and label each volcano.

© Earth and Space Science • Volcanoes

Name: _____   Date: _____

Volcano Zone   WS 3.3.2A

The following chart contains a list of volcanoes. Plot each volcano on the map. Research the most recent eruption of each volcano and write a brief summary about the eruption.

Volcanoes

Name of Volcano	Location	Description of Eruption
Ruapehu, New Zealand	38°S, 176°E	
Tungurahua, Ecuador	1°S, 78°W	
Mount Pinatubo, Philippines	15°N, 120°E	
Arenal, Costa Rica	10°N, 85°W	
Krakatau, Indonesia	6°S, 105°E	
Popocatepétl, Mexico	19°N, 98°W	
Mount Oyama, Japan	34°N, 139°E	
Mount Cleveland, Alaska	52°N, 169°W	
Klyuchevskaya Sopka, Russia	56°N, 160°E	
Mount Etna, Italy	37°N, 15°E	
Mount Saint Helens, USA	46°N, 122°W	

1. Which volcanoes do you believe are formed in a subduction zone? Why? _____
   _____
   _____

2. Which volcanoes do you believe are formed from an oceanic-oceanic plate collision? Why? _____
   _____

3. Which volcano is not part of the Ring of Fire? _____

4. What type of plate interaction do you believe created this volcano? Why? _____
   _____
   _____
   _____

© Earth and Space Science • Volcanoes

Name: _____ Date: _____

Volcano Zone

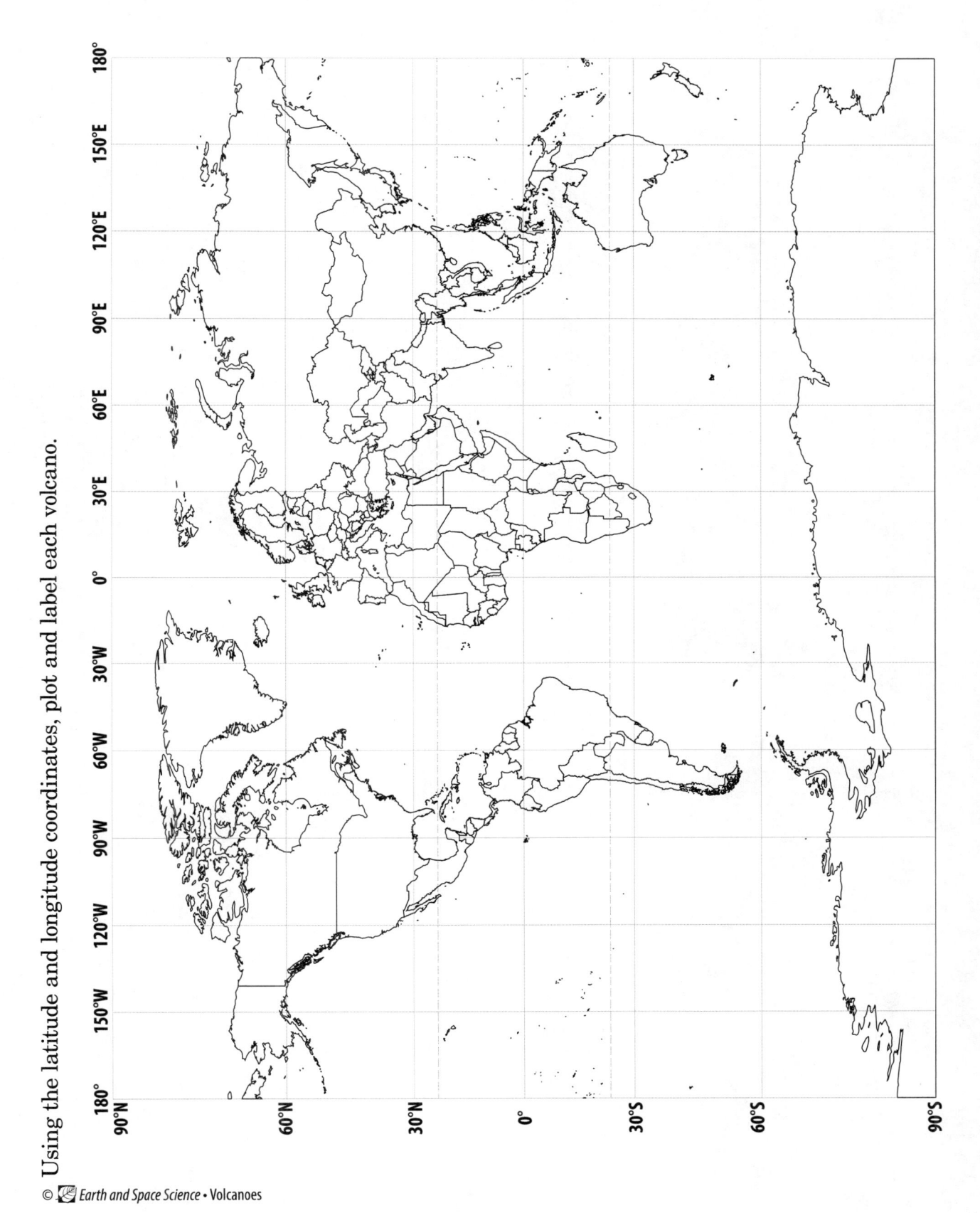

WS 3.3.2A

Using the latitude and longitude coordinates, plot and label each volcano.

© Earth and Space Science • Volcanoes

Name: _____   Date: _____

Comparing Eruptions   WS 3.3.4A

Compare the components of each eruption. Look for a relationship between the characteristics of the eruption and the level of violence.

Comparing Eruptions

Eruption Type	Steam/Gas Content	Lava Flow	Pyroclastic Flow	Ash Cloud	Level of Violence
Icelandic					
Hawaiian					
Strombolian					
Vulcanian					
Peléan					
Plinian					

1. Which eruptions have the smallest gas content? _____
_____

2. Which eruptions have the greatest gas content? _____
_____

3. Which eruptions have runny lava? _____
_____

4. Which eruptions have thick lava? _____
_____

5. What do you think causes eruptions to be more violent? _____
_____
_____

© Earth and Space Science • Volcanoes

Name: _____ Date: _____

Artistic Interpretations WS 3.3.4B

Research paintings that depict volcanic eruptions. Choose one to research fully and answer the questions below.

1. Who is the artist? What country is the artist from? _____
_____

2. Does the painting depict a specific eruption? If yes, which one? _____
_____

3. When was the painting completed? _____

4. What do you like about the painting? _____
_____
_____
_____

5. What do you think the artist feels about the eruption? Why? _____
_____
_____
_____

6. Create your own artwork of a volcanic eruption.

© Earth and Space Science • Volcanoes

Name: _____ Date: _____

Lab 4.1.1A The Dissolving Power of Water

QUESTION: Does the addition of sugar (solute) affect the volume of water (solvent)?

HYPOTHESIS: _____

_____

EXPERIMENT:

You will need:	• wax marking pencil	• electronic balance
• 100 mL graduated cylinder	• spoon	• metric ruler
• two 100 mL beakers	• 50 g sugar	

Steps:
1. Measure 50 mL of hot water and place it in a beaker. Measure the height of the water in centimeters.

    • Exact volume of water (mL): _____

    • Height of water (cm): _____
2. Add 16 g of sugar (about 4 level teaspoons) to the empty beaker. Measure the height of the sugar in centimeters.

    • Height of sugar (cm): _____
3. Add 16 g of sugar (about 4 level teaspoons) to the beaker containing water, stirring vigorously.
4. Measure the height of the water and sugar mixture in centimeters.

    • Exact volume of mixture (mL): _____

    • Height of mixture (cm): _____

ANALYZE AND CONCLUDE:
1. Calculate the volume difference between the mixture and the water.

    Volume of Mixture – Volume of Water =   _____

2. Calculate the difference in height between the mixture and the water.

    Height of Mixture – Height of Water =   _____

3. Compare the height of the sugar and the water height difference. Did the water rise as high as expected after the sugar was added? Explain the height variations.

_____

_____

_____

_____

© Earth and Space Science • Water

Lab 4.1.1A The Dissolving Power of Water

4. What does this experiment tell you about the molecular structure of sugar?

Explain.

5. Design and test an experiment using other liquids or other solids. Receive approval for the proposed experiment from the teacher before proceeding. Explain the test results for the other liquids or solids tested with regard to polarity.

Name: _____   Date: _____

Lab 4.1.2A Cloud Capacity

QUESTION: How much water will a standard cotton ball hold?

HYPOTHESIS: _____

_____

EXPERIMENT:

You will need:	• scale	• 100 mL beaker
• cotton ball	• eyedropper	

Steps:
1. Fill the beaker with 50 mL of water.
2. Separately measure and record the mass of the cotton ball and one drop of water.

   • Mass of cotton ball: _____ grams

   • Mass of one drop of water: _____ grams
3. Predict how many drops of water can be added to the cotton ball before the water begins to drip.

   • Estimated number of water drops: _____ drops
4. Hold a small section of a cotton ball by pinching it with your index finger and thumb. The cotton ball should be held above the beaker.
5. Apply drops of water, one drop at a time, to the cotton ball.
6. Count how many drops of water are added to the cotton ball. Stop counting the number of drops added as soon as the cotton ball begins to drip.

   • Total number of water drops added: _____ drops
7. Record the number of drops that other groups measured.

_____

ANALYZE AND CONCLUDE:

1. Was the estimated number of water drops equal to the actual number of water drops added to the cotton ball? _____

2. Did each group record the same amount of water drops added to the cotton ball? Why did the results vary? _____

_____

_____

_____

_____

Lab 4.1.2A Cloud Capacity

3. Graph the results on the chart below. Add a title for the graph.

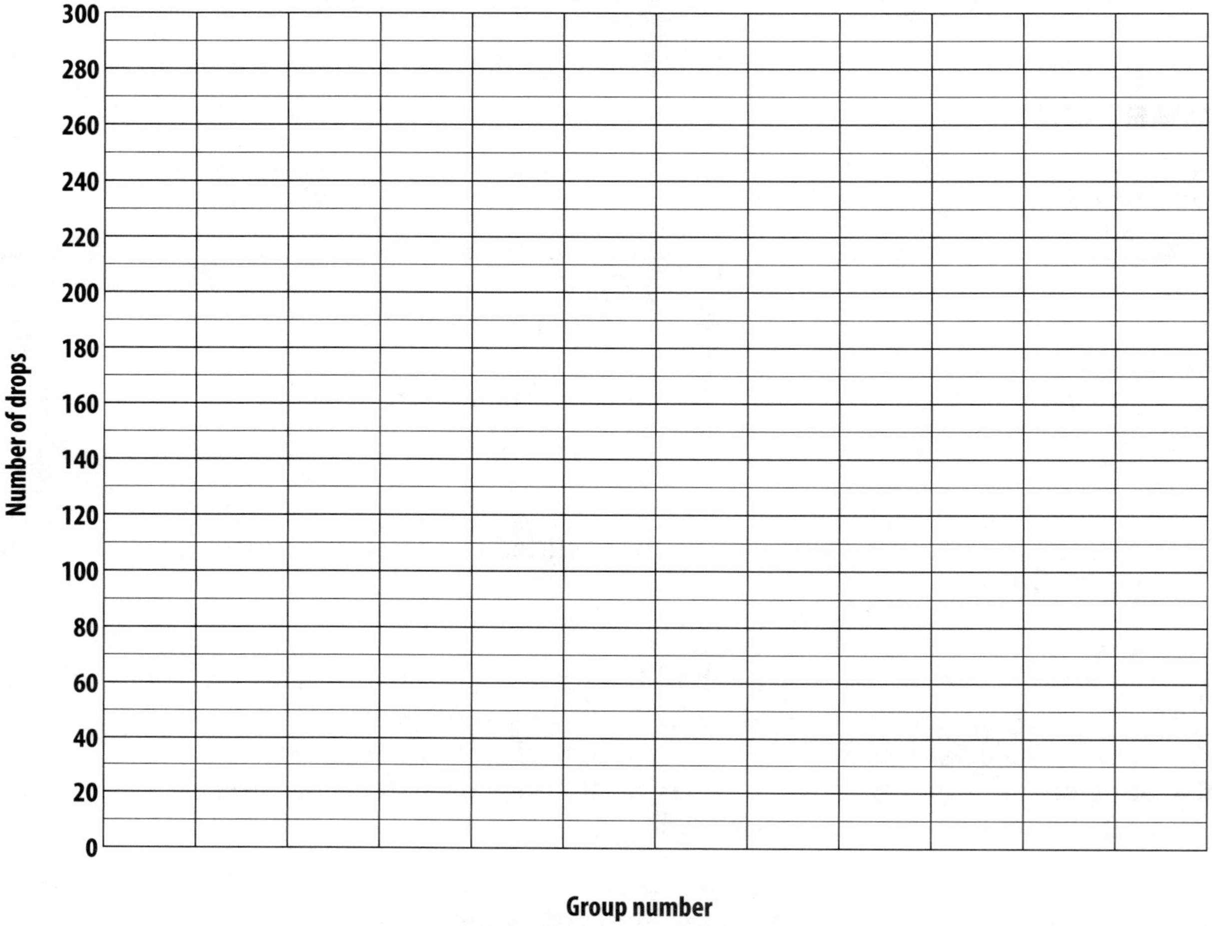

4. What forces allow the cotton ball to hold such a tremendous amount of water?

5. Relate the saturation of the cotton ball to the saturation of a cloud.

Name: _____ Date: _____

Lab 4.1.4A Groundwater Contamination

QUESTION: How do different sediments affect water permeability?

HYPOTHESIS: _____

_____

EXPERIMENT:

You will need:	• aquarium pebbles	• metric ruler
• 25 cm × 16 cm section of screen	• sand	• timer
• 500 mL beaker	• topsoil	
• 250 mL beaker	• food coloring	

Steps:
1. Roll the screen to form a long cylinder with a 5 cm diameter. The cylinder represents a well. Place and hold the cylinder in the center of a 500 mL beaker.
2. Add aquarium pebbles to the bottom of the beaker to form a 4 cm layer.
3. Pour water over the pebbles until the water level measures about 2 cm. This water represents groundwater.
4. Add a 2 cm layer of sand above the pebbles.
5. Add a 2 cm layer of topsoil above the sand. Carefully release the cylinder.
6. Add a few drops of food coloring to 150 mL of water in a separate beaker. The colored water represents contaminated water.
7. Pour 100 mL of the contaminated water over the topsoil around the well.
8. Record the amount of time it takes for the contaminated water to enter the groundwater. _____ minutes:seconds.
9. Record the amount of time it takes for the contaminated water to enter the well. _____ minutes:seconds.

ANALYZE AND CONCLUDE:

1. Did the contaminated water travel through each layer at the same rate of speed? Why? In which sediment layer did the water travel the fastest? the slowest?

_____

_____

2. Was the color of the well water the same color as the original contaminated water? Explain the differences. _____

_____

3. What type of improvements could be made to prevent contamination of the well water? _____

_____

© Earth and Space Science • Water

Name: _____ Date: _____

Lab 4.1.5A Hard Water

QUESTION: Does hard water affect the cleansing power of soap?

HYPOTHESIS: _____

_____

EXPERIMENT:

You will need:	• 100 mL distilled water	• metric ruler
• 2 plastic bottles with lids	• 20 g Epsom salts	
• 100 mL graduated cylinder	• dishwashing liquid	

Steps:
1. Pour 50 mL of distilled water into each plastic bottle.
2. Add 20 g of Epsom salts to one of the plastic bottles. Cover the bottle with the lid and gently shake it until all of the salt is dissolved.
3. Place a small amount of dishwashing liquid into both plastic bottles.
4. Cover both bottles with lids and shake them vigorously to produce bubbles.
5. Measure and record the height of the bubbles.

- Height of distilled water bubbles: _____ cm
- Height of Epsom salt mixture bubbles: _____ cm

ANALYZE AND CONCLUDE:

1. Were the bubble heights the same? Explain any differences. _____

_____

_____

_____

_____

2. What made the hard water "hard"? _____

_____

3. Will the hard water clean as well as standard or soft water? Why? _____

_____

_____

_____

_____

© Earth and Space Science • Water

Name: _____ Date: _____

Lab 4.1.7A Pond Exploration

QUESTION: How does the surrounding land contribute to the health of a pond-water system?

HYPOTHESIS: _____
_____

EXPERIMENT:

You will need:	• 2 water hardness test strips	• tweezers
• tape measure	• 2 litmus paper strips	• ice cube tray
• jar	• 2 ammonia test strips	• hand lenses
• thermometer	• 2 nitrate test strips	• pipette
• vial	• sieve	• D-frame aquatic dip net
• dissolved oxygen test kit	• bucket	

Steps:
Site Description

1. Topography surrounding the pond: _____
_____
_____
_____

2. Current latitude/longitude: _____

3. Site length (meters): _____

4. Minimum pond width (meters): _____

5. Maximum pond width (meters), if measurable: _____

6. Number of transects: _____

7. What is the dominant vegetation type in the area (none, cultivated, meadow, scrub, or forest)? Is the vegetation located on both sides of the pond? _____
_____

8. Is the dominant substrate silt/sand (<2 mm), pebbles (2–8 mm), gravel (8–64 mm), cobblestone (64–256 mm), or boulders (>256 mm)? _____
_____

9. What is the estimated overhead forest cover (none, 1%–25%, 26%–50%, 51%–75%, or 76%–100%)? _____
_____

© *Earth and Space Science • Water*

Lab 4.1.7A Pond Exploration

10. Describe the weather conditions, any notable or unusual site conditions, sampling problems, and sightings or observations—including plants, animals, invasive species, and human activities.

   _____
   _____
   _____
   _____
   _____
   _____
   _____

11. Identify and describe any potential sources of pollution seen on or near the site.

   _____
   _____
   _____
   _____
   _____

12. Sketch all pond features, transects, vegetation, and nearby permanent features, including roads, buildings, paths, and bridges. Label features according to a map legend and indicate directions.

Name: _____ Date: _____

Lab 4.1.7A Pond Exploration continued

Turbidity
Turbidity measures the cloudiness of water. The greater the turbidity of the water, the less life it can support. Water becomes turbid, or cloudy, when the suspended solids in the water increase. Soil erosion, urban runoff, waste discharge, algal blooms, and pond substrate disturbance all contribute to increased turbidity.
1. Place the jar with a sticker horizontally in the water and gently sweep the jar from side to side for 15 seconds.
2. Once the jar is full of water, remove it from the pond, and look through the top to observe the sticker on the bottom of the jar. Is the sticker 100% visible, 50% visible, or 0% visible? _____

Temperature
Different species thrive at different water temperatures. If the temperature changes in a short amount of time, animals can experience stress, and decreased oxygen levels may be present. A temperature of 13°C is ideal during the fall and spring months.
1. Place a jar in the pond water at least 10 cm below the surface.
2. After 1 minute has elapsed, remove the jar and use a thermometer to record the temperature of the water in the jar. _____ °C

Dissolved Oxygen
Ponds with higher amounts of dissolved oxygen support more life. Oxygen levels are reduced when water becomes polluted with fertilizers, sewage, animal feces, and garden waste, which use oxygen to decompose.
1. Submerge a jar in the pond water for 20 seconds.
2. Withdraw the jar from the pond and use it to completely fill a small vial.
3. Take 2 oxygen tablets from the dissolved oxygen test kit, drop them into the vial, and place the lid on the vial. Water should overflow out of the vial.
4. Invert the vial for about 5 minutes until the oxygen tablets are fully dissolved.
5. The water color should change 5 minutes after the tablets have fully dissolved.
6. Follow the directions in the oxygen test kit to record the amount of oxygen dissolved in the water sample. _____ ppm

Water Hardness
A water hardness test measures the amount of calcium and magnesium present. If the level is high, this indicates there is too much algae. If the level is low, very little algae is present. To raise the calcium and magnesium levels, different salt solutions (not table salt) can be added to the water. A normal water hardness level ranges between 150 and 330 ppm.
1. Fill a vial with pond water.
2. Dip a water hardness test strip in the water and observe the color change.
3. Compare the strip's color to the key on the water hardness test strip container and record the value. _____ ppm

© Earth and Space Science • Water

Lab 4.1.7A Pond Exploration

pH
Most aquatic animals thrive in a water pH range of 6.5–8.0. Acid rain and waste water discharge can affect the natural pH of a water source.
1. Fill a vial with pond water.
2. Dip a litmus test strip in the water and observe the color change.
3. Compare the strip's color to the pH key on the litmus test strip container and record the pH level. _____

Ammonia
Decaying organic material and excess amounts of fish waste can contribute to a high ammonia level in pond water. If the ammonia level is too high, different types of bacteria can be added to the water. An ideal ammonia reading is <0.1 ppm.
1. Fill a vial with pond water.
2. Dip an ammonia test strip in the water and observe the color change.
3. Compare the strip's color to the ammonia key on the ammonia test strip container and record the value. _____ ppm

Nitrates
High levels of nitrates can be observed in water that contains a large quantity of plant life or algae. If the nitrate level is too high, buffers can be created between the water source and the pollution source, along with limiting the use of fertilizers. An ideal nitrate level is <20 ppm.
1. Fill a vial with pond water.
2. Dip a nitrate test strip in the water and observe the color change.
3. Compare the strip's color to the nitrate key on the nitrate test strip container and record the value. _____ ppm

Organisms
1. Use the jar to collect a water sample from just below the water surface.
2. Pour the water through a sieve and into a bucket.
3. Use tweezers to place any organisms into compartments in an ice cube tray. Use a pipette to add a little water to each compartment to keep the organisms moist.
4. Use a pipette to gather small organisms from the bucket that passed through the sieve and add them to the ice cube tray.
5. Examine the organisms with a hand lens and draw them in the space provided.
6. Return the organisms to the pond.
7. Repeat Steps 1–6 with water drawn from halfway between the surface of the pond and the bottom of the pond.
8. Repeat Steps 1–6 with water drawn from near the bottom of the pond.
9. Using a D-frame aquatic dip net, make several slow back-and-forth sweeps through the pond.
10. Use the space provided to draw and label the organisms in the net.

Name: _____ Date: _____

Lab 4.1.7A Pond Exploration continued

Surface Organisms Drawings

Middle Organisms Drawings

Bottom Organisms Drawings

© *Earth and Space Science • Water*

Lab 4.1.7A Pond Exploration

ANALYZE AND CONCLUDE:

1. Using the data, is the overall health of the pond good, fair, or poor? Explain.

_____
_____
_____
_____
_____

2. What factors contributed to the health of the pond? _____
_____
_____
_____
_____

3. How could the health of the pond be improved? _____
_____
_____
_____
_____

4. What kind of impact would high turbidity have on the animals and plants in the pond habitat? _____
_____
_____
_____

5. What kind of impact would the removal of vegetation around the water have on the pond habitat? _____
_____
_____
_____
_____

Name: _____ Date: _____

Physical Properties of Water   WS 4.1.1A

Complete each station as directed by your teacher.

Surface Tension
1. Place the wire screen over the opening of the canning jar and fill the jar with water just to overflowing.

2. Cover the wire screen with the card stock and invert the jar.

3. Slowly slide the card stock off to the side.

4. Observations: _____
_____

5. Why does the water remain in the inverted jar? _____
_____

6. On a separate piece of paper, design a surface tension experiment using different jar sizes and covers.

Float It
1. Fill a bowl with water.
2. Place a paper clip on a small square of paper.
3. Gently lay the square of paper with the paper clip on the water's surface.

4. Observations: _____
_____

5. Why does the paper clip float on top of the water? _____
_____

6. Add a drop of detergent.

7. Explain what happened to the paper clip after the detergent was added. _____
_____
_____

Drop It
1. Use the eyedropper to place one drop of water on a small sheet of waxed paper.
2. Draw the side view of the drop of water.

© Earth and Space Science • Water

Physical Properties of Water

WS 4.1.1A

3. Draw the top view of the drop of water.

4. Place one drop of isopropyl alcohol on a small sheet of waxed paper.

5. Draw the side view of the drop of isopropyl alcohol.

6. Draw the top view of the drop of isopropyl alcohol.

7. How do the drops of water and isopropyl alcohol compare? _____
_____
_____

Climbing Water

1. Place one end of a drinking straw in a bowl of water.

2. Use a hand lens to examine the water in the straw. Record your observations.
_____
_____

3. Why does the water climb up the straw? _____
_____

Name: _____ Date: _____

Physical Properties of Water continued   WS 4.1.1A

4. How can this property help living things? _____
_____
_____

5. Place one or two capillary tubes in the water.

6. How does the width of the tube relate to how high the water climbs? _____
_____
_____

Heat Retention
1. Use a hot plate to slowly heat about 75 mL of water in a beaker.
2. In the table below, record the temperature each minute until the water reaches its boiling point. Indicate when small bubbles begin to form by placing an arrow to the side of the table. Record the temperature 4 minutes after it has reached the boiling point. Remove beaker from heat.

Heat Retention

Time (min)	Temperature (°C)

3. Explain why so much time elapsed before the water boiled. _____
_____
_____

© Earth and Space Science • Water

Name: _____ Date: _____

Amount of Water   WS 4.1.2A

Is it better to take a shower or a bath? Should people brush their teeth with the water running or turn the water off? Review the table below and decide.

Activity	Gallons Used	Liters Used
Pre-1993 showerhead	3–8 per minute	11.4–30.3 per minute
Standard showerhead	2.5 per minute	9.5 per minute
Low-flow showerhead	1.5 per minute	5.7 per minute
Bathtub, full	30–45	113.6–170.3
Bathtub, one-fourth to one-third full	7.5–15	28.4–56.8
Brushing teeth, faucet running	2.2 per minute	8.3 per minute
Brushing teeth, faucet off	0.25–0.5	0.95–1.9
Conventional toilet	3.5–5 per flush	13.2–18.9 per flush
Low-flow toilet	1.6 per flush	6.1 per flush
High-efficiency toilet	1.28 per flush	4.85 per flush
Hand washing, standard aerator	2.2 per minute	8.3 per minute
Hand washing, low-flow aerator	0.5 per minute	1.9 per minute
Washing machine, conventional top-loader	35–50 per load	132.5–189.3 per load
Washing machine, front-load washer	13–20 per load	49.2–75.7 per load
Dishwasher, standard cycle	10–14	37.9–52.9
Dishwasher, water-conserving model	4.5–7	17–26.5
Dishwashing by hand, full basin/wash and rinse	2–4	7.57–15.1
Dishwashing by hand, running water	2.2 per minute	8.3 per minute
Washing car with bucket	10	37.9
Washing car with hose	6.4 per minute	24.2 per minute
Watering yard with sprinkler	140 per hour	529.9 per hour

1. Calculate how much water you use in one day. Refer to the table above. _____

2. Calculate how much water you use in one month. Refer to the table above. _____

3. In what ways could you reduce the amount of water used? _____

© Earth and Space Science • Water

Name: _____   Date: _____

Comparing Water Budgets   WS 4.1.2B

Not all regional water budgets are the same. Sometimes more water falls on an area than evaporates. Plant transpiration rates are also affected by environmental factors, such as wind, temperature, sunlight availability and intensity, humidity, soil type, precipitation, and land slope. Examine the charts below that depict annual precipitation compared to potential transpiration for Town A and Town B and answer the questions below.

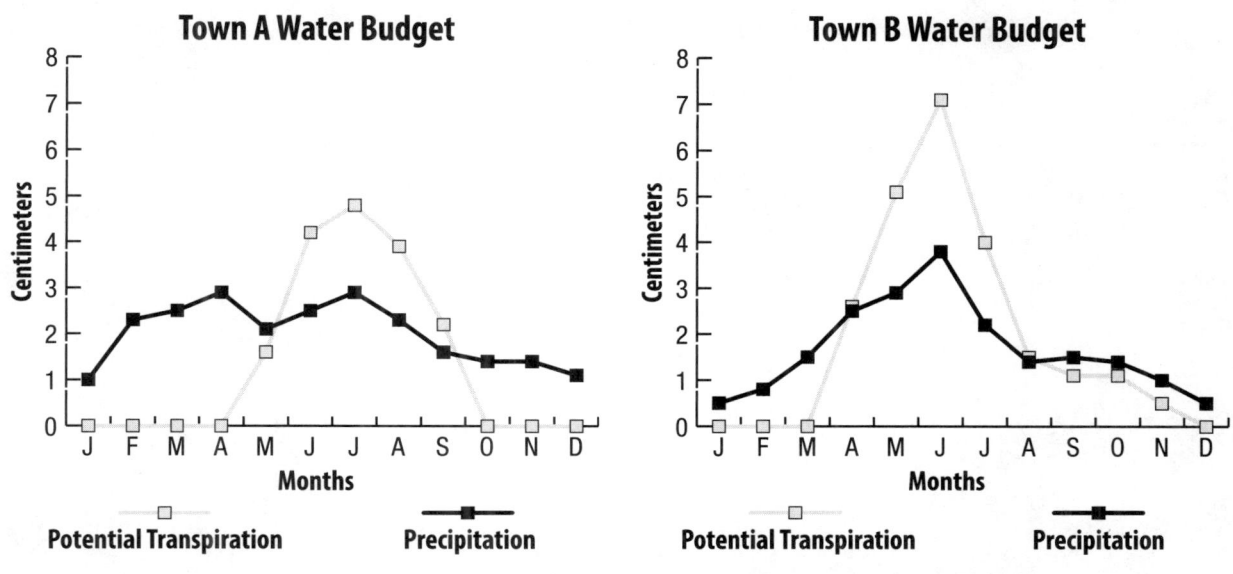

Town A Water Budget

Town B Water Budget

Potential Transpiration   Precipitation

1. During what months does Town B have a water surplus? Which month indicates the highest water surplus? _____

2. During what months does Town B have a water shortage? Which month indicates the greatest water shortage? _____

3. Does Town A or Town B have an overall water shortage for the year? Which town has an overall water surplus for the year? _____

4. During what months does Town A not have any potential transpiration? Explain. _____

5. Town B is located downstream from Town A. What would you predict would happen if Town A had a very large water surplus? What kind of problems might exist with a very large water surplus? Suggest ways that scientists could minimize these problems. _____

© Earth and Space Science • Water

Name: _____   Date: _____

Global Glaciers
WS 4.1.3A

Glaciers can be found on every continent except Australia. Color the locations of all of the glaciers throughout the world. Draw a star on your current geographic location.

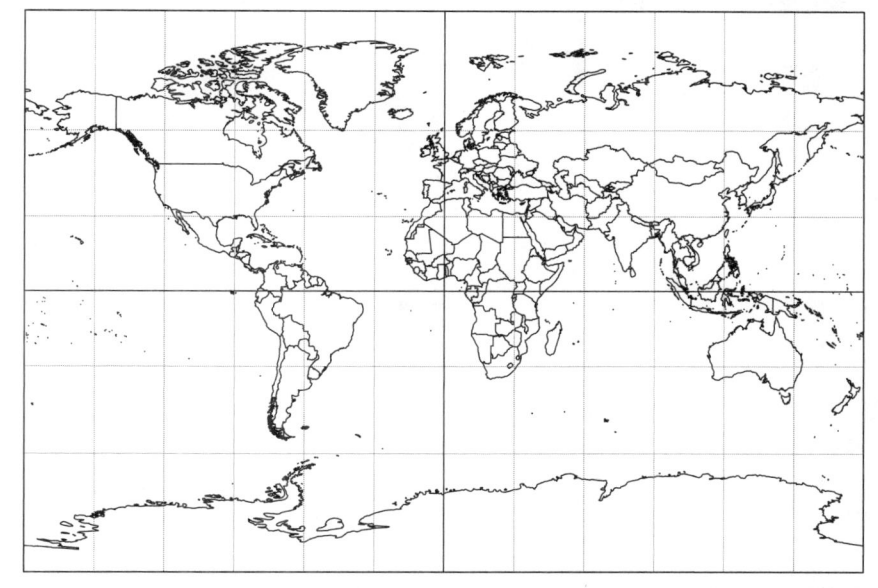

Imagine you are a glacier and the only way for you to survive is by eating snow. You would want to avoid intense sunlight and heat as they will cause you to shrink.

1. Where would you live? Locations to choose include polar regions, tropical regions at high elevations, and between the Tropic of Cancer and the Tropic of Capricorn at moderate elevations. Explain why you chose a particular region.

   _____
   _____

2. Using the climate of your location, how much snow would you be able to eat? Also, how frequently would you be able to eat snow? _____
   _____
   _____

3. How would you avoid the intensity of the sun? _____
   _____

4. Where are some of the largest mountain glaciers located throughout the world?
   _____

5. Why are these glaciers so large? Research and include in your answer the average temperature and amount of precipitation on the glacier. _____
   _____
   _____

© Earth and Space Science • Water

Name: _____ Date: _____

Glaciers in the Water Cycle   WS 4.1.3B

Use your Student Edition to fill in how glaciers fit into the water cycle.

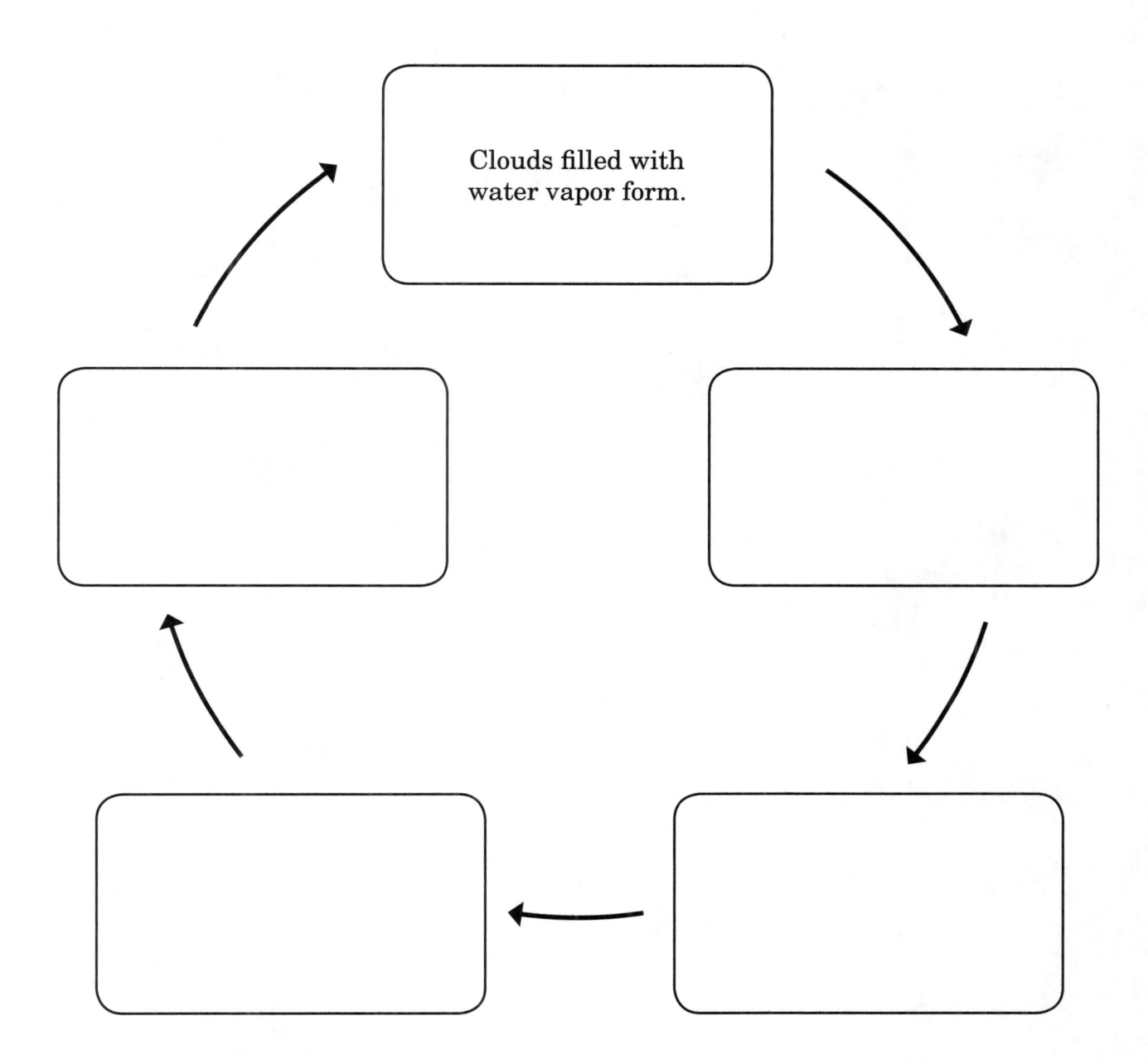

Clouds filled with water vapor form.

Name: _____ Date: _____

River Creation
WS 4.1.6A

Perform the following activities with your group to determine what type of river is being created. Answer the questions below.

Fill a long, plastic tray with sand several centimeters deep. Place four or five bricks under one end of the tray to create a steep incline. Position the other end of the tray on the edge of a sink to ensure the water will flow into the sink. Line the sink with paper towels to catch any sand that washes out of the tray. Connect one end of the hose to a faucet, place the other end of the hose at the elevated end of the tray, and increase the water flow to represent fast-moving water.

1. What type of channel is produced? _____

2. What type of river is this? _____

Smooth the sand and decrease the incline by removing one or two bricks. Carve a snakelike groove in the sand. Position the hose so water flows into the groove. Lower the water flow to produce a meandering erosion pattern.

3. What characteristics do you observe? _____

_____

4. What type of river is this? _____

Smooth the sand and set the tray on one brick to create a low incline. Reduce the water flow to produce large meanders.

5. What do you observe in the channel? _____

_____

6. What factors have changed since you created the first river in the tray? _____

_____

7. What type of river is this? _____

8. Explain how you would create a rejuvenated river. _____

_____

_____

© Earth and Space Science • Water

Name: _____ Date: _____

Dams
WS 4.1.8A

Answer the following questions using information presented by the tour guide or found on the Internet.

1. How large is the dam (in cubic meters)? _____

2. How tall is the dam (in meters)? _____

3. What is the water flow rate in cubic meters per second? _____

4. How many dams are on the river? _____

5. What percentage of the area's electrical energy does the dam supply? _____
   _____

6. What is the cost of a kilowatt hour of electricity supplied by the power plant?
   _____

7. What is the cost of a kilowatt hour of electricity supplied by a nearby fossil fuel or nuclear power plant? _____
   _____

8. How much does it cost every year to maintain the power plant? _____
   _____

9. How do the costs compare with power plants that use other energy sources?
   _____
   _____

10. What types of pollution or environmental problems does the power plant produce? _____
    _____

11. What steps have been taken to solve environmental problems, such as helping aquatic wildlife? _____
    _____
    _____

12. How has the power plant affected river life? _____
    _____
    _____
    _____

© Earth and Space Science • Water

Name: _____ Date: _____

Lab 4.2.1A Ocean Water Density

QUESTION: What things affect the density of ocean water?

HYPOTHESIS: _____

_____

EXPERIMENT:

You will need:	• distilled water	• bucket
• fine-point permanent marker	• 250 mL beaker	• ice
• metric ruler	• thermometer	• 40 g of salt
• 3 drinking straws	• clock	• colored pencils
• modeling clay	• metal beads	

Steps:
1. Use a permanent marker to mark 5 mm increments along the drinking straws.
2. Label the first straw *1*, the second straw *2*, and the last straw *3*.
3. Roll three small pieces of clay into balls, each about 1 cm in diameter. Plug one end of each straw with a clay ball. Use the smallest amount of clay possible that will completely seal the end of the straw. If the straws contain too much clay, they will not float.
4. Pour 200 mL of distilled water at room temperature into a 250 mL beaker. All measurements taken with this water will be recorded next to Warm Freshwater on the data table.
5. Place the thermometer in the beaker, allow it to sit for 2 minutes, and record the temperature on the data table. Remove the thermometer.
6. Place Straw 1 in the water with the plugged end down. If it does not stay upright in the water, drop some metal beads into it, one at a time, until the straw stays upright. The straw must float in the beaker without touching its bottom or sides.
7. Calculate the height of the straw's submerged portion by counting the number of underwater marks. If the water level falls between two lines, estimate the measurement. Remember to take the measurement from the bottom of the meniscus. Multiply the number of lines by 5 to calculate the total number of millimeters and round to the nearest whole number. Record the measurement on the data table. For example, if 3.25 lines are underwater, multiply by 5 mm to obtain 16.25 mm, and round to 16 mm.
8. Repeat Steps 6–7 with Straw 2 and Straw 3.
9. Average the measurements of all three straws and record it on the data table.
10. Place a thermometer into the beaker of water and place the beaker in a bucket of ice until the water temperature is 4°C–7°C. Remove the beaker from the ice and repeat Steps 6–9, recording the data next to Cold Freshwater on the data table. Do not change the number of metal beads in the straws.
11. Prepare a solution of 40 g of salt in 200 mL distilled water at room temperature.
12. Repeat the above procedures, recording the data under Warm Saltwater and Cold Saltwater.

© Earth and Space Science • Oceans

Lab 4.2.1A Ocean Water Density

	Temperature	Straw 1 Height	Straw 2 Height	Straw 3 Height	Average Height
Warm Freshwater (WFW)					
Cold Freshwater (CFW)					
Warm Saltwater (WSW)					
Cold Saltwater (CSW)					

13. Create a line graph in the space provided to plot the table's information. Plot the temperature along the *x*-axis, and plot the average height along the *y*-axis. Use one color for the freshwater and a different color for the saltwater. Use the acronyms to label the points.

ANALYZE AND CONCLUDE:

1. How can the submerged straw indicate the density of the water sample? _____

_____

_____

2. Which sample of water had the greatest density? Which had the least? _____

_____

_____

3. How did temperature affect the density of each water sample? _____

_____

4. Where in the ocean would you expect to find water that is least dense? _____

_____

_____

5. How could you create water that is more dense? _____

_____

_____

6. Why were three different straws used for each density measurement? _____

_____

_____

_____

Name: _____ Date: _____

Lab 4.2.2A Wind and Waves

QUESTION: How are waves and surface currents created and sustained?

HYPOTHESIS: _____
_____

EXPERIMENT:

You will need:	• bendable straw	• dry-erase marker
• clear, deep plastic container	• tape	• 4 different colors of food coloring
• 3 bricks	• string	
• sand	• metric ruler	

Steps:
1. Label the two ends of the container A and B.
2. Lay a brick against one corner of end A so the brick's long end is parallel to the container's short end. Lay the second brick so one end is on the first brick and the other end is on the bottom of the container. The two bricks should look like a T with a big bump in the middle.
3. Add sand to the container until the level of sand is 3 cm from the top. Keeping the bricks in place, scrape the sand so most of it is mounded over the bricks at end A.
4. Bend the straw into an L shape.
5. Tape the straw to the inside of end B so it is centered between the corners and 1 cm down from the top edge. The short leg of the L should point at end A, and the straw's long leg should point straight up.
6. Tie the string around the third brick and lay the brick on its long edge parallel to end B. Drape the string over the side of the container and add water to the container until it is just below the straw.
7. Wait until there are no ripples in the water.
8. Measure the water level to the nearest millimeter from the tabletop.
9. One partner should blow very gently through the straw for 10 seconds to simulate wind as the other partner looks into the container at end A and uses a dry-erase marker to mark the maximum wave height on the nonsandy side.
10. Measure the wave height from the tabletop to the nearest millimeter. Find the amplitude by subtracting the calm water level from the wave height.
11. Wait until there are no ripples in the water.
12. Place 1 drop of food coloring in the water at each of the container's corners so each corner has a different color.
13. Repeat Steps 9–10, but create wind for 2 minutes without changing the force of blowing. Mark and measure the wave height every 30 seconds, but do not stop blowing. Record the amplitude for each wave measurement.
14. Wait until there are no ripples in the water.
15. Remove the straw, then quickly pull the string to flip the third brick over to simulate a tsunami.

© Earth and Space Science • Oceans

Lab 4.2.2A Wind and Waves

ANALYZE AND CONCLUDE:

1. Look at the sand. How might waves affect a beach over many years? _____
_____
_____
_____

2. Examine the Wave Data Table and explain the results. _____
_____
_____
_____

3. Draw the currents that formed during the 2 minutes of blowing. Show where the colors traveled.

4. Where did the tallest waves form? _____
_____

5. Where did the shortest waves form? _____
_____

6. How did the tsunami wave compare with the waves created by wind? _____
_____
_____

Name: _____ Date: _____

Lab 4.2.3A Tides

QUESTION: How are the tides on Earth affected by the position of the moon?

HYPOTHESIS: _____

_____

EXPERIMENT:

You will need:	• compass or round object	• metric ruler
• tape	• 3 objects with equal masses	
• 11" × 17" piece of paper	• 2 springs	

Steps:
1. Tape the paper down on a flat surface.
2. Draw a circle on the piece of paper using the compass or round object. The circle should have at least a 25 cm diameter. This image represents the earth.
3. Label the three objects *A*, *B*, and *C*.
4. Attach A to B with one spring and attach B to C with the other spring.
5. Arrange the objects on the piece of paper with B in the middle of the circle, A touching the line just outside of the circle to the left of B, and C touching the line just outside of the circle to the right of B. Draw a small dot on one side of each of the objects and label the dot in relation to the objects.
6. Slide A to the left and observe the movement of B and C.
7. Draw a small dot on one side of each of the objects in the new position and label the dot in relation to the objects.
8. Measure the distance each object moved.

   - Distance B moved: _____ cm
   - Distance C moved: _____ cm

ANALYZE AND CONCLUDE:

1. Describe the movement of B and C when A is pulled. _____

_____

_____

2. Do B and C move equal distances? Why or why not? _____

_____

_____

3. If A represents the moon's gravitational force and the circle represents Earth, what do B and C represent? _____

Lab 4.2.3A Tides

4. If another object (D) with a weak gravitational force is placed in this configuration directly above B, how will the movement of B and C be affected? _____
_____
_____

5. What type of tides would be produced in the configuration described in question 4?
_____

Name: _____ Date: _____

Lab 4.2.4A Predicting Ocean Floor Topography

QUESTION: Is the line and sinker method an accurate way to determine the appearance of the ocean floor?

HYPOTHESIS: _____
_____

EXPERIMENT:

You will need:	• mystery box	• metric ruler	• probe

Steps:
1. Do not remove the lid from the mystery box.
2. Starting at one end of the box, measure the distance in centimeters between the edge of the box and each hole. Record measurements in the table below.

Hole Identification	Distance (cm)	Hole Identification	Distance (cm)

3. Insert the probe into each hole. When the probe hits a surface, place a finger where the probe meets the box, remove the probe, and measure the length of the probe from the tip to your finger. Record measurements in the table below.

Hole Identification	Distance (cm)	Hole Identification	Distance (cm)

Lab 4.2.4A Predicting Ocean Floor Topography

4. Plot the distance (*x*-axis) versus height (*y*-axis) to create a line graph below and connect the points. Label and title the graph.

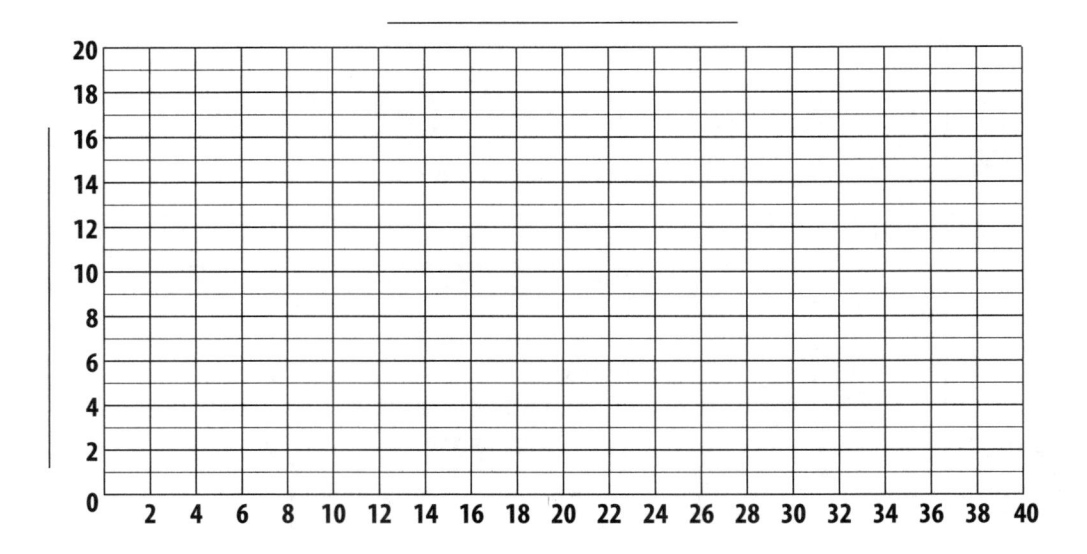

ANALYZE AND CONCLUDE:

1. Relate the image of the line to the actual box floor features. _____

2. In the space below, draw a hypothesized view of the box floor. Imagine that the box is the ocean floor. Label the regions that represent the continental shelf, continental slope, and oceanic ridge. _____

3. Remove the box lid. Does the box floor look identical to the line graph? Why?

4. Using the lab results, is the line and sinker method the best way to map the floor of the box? What other methods might an oceanographer use when mapping the ocean floor? _____

© Earth and Space Science • Oceans

Name: _____ Date: _____

Lab 4.2.6A Distillation

QUESTION: What method of distillation most efficiently separates salt from water?

HYPOTHESIS: _____
_____

EXPERIMENT:

You will need:	• distilled water	• metric ruler
• boiling chips	• rubber tubing	• Bunsen burner or hot plate
• 100 mL saltwater	• rubber stopper with hole	• test tube
• 250 mL Erlenmeyer flask	• clamp	• two 600 mL beakers
• conductivity probe	• ring stand	• ice

Steps:
1. Place a few boiling chips and 100 mL of saltwater in the Erlenmeyer flask.
2. Set the conductivity probe to the highest setting using mg/L units.
3. Rinse the probe with distilled water.
4. Place the probe in the saltwater solution and record the conductivity reading.
_____ mg/L
5. Rinse the probe again with distilled water.
6. Place the rubber tubing in the hole of the rubber stopper and insert the stopper into the mouth of the Erlenmeyer flask.
7. Position the Erlenmeyer flask in the clamp on the ring stand at least 2.5 cm above the Bunsen burner or place the flask directly on the hot plate. See illustration for proper set up.
8. Extend the rubber tubing to fit inside the test tube. Do not allow the rubber tubing to form any kinks; this could cause a potentially dangerous pressure problem.
9. Fill one 600 mL beaker with ice and water to create an ice-water bath.
10. Place the test tube in the ice-water bath.
11. Heat the saltwater inside the Erlenmeyer flask just to boiling. Do not allow the boiling water to reach the rubber stopper. If it looks like the bubbles may reach the rubber stopper, remove the heat source and adjust the burner settings.
12. Observe the desalinated water collecting in the test tube.
13. When most of the saltwater has boiled off, turn off the heat source and allow the apparatus to cool for 10 minutes. Do not allow all of the saltwater to evaporate.
14. Fill the second 600 mL beaker with room-temperature water and carefully transfer the test tube from the ice water to the room-temperature water.
15. Adjust the conductivity probe to the lowest setting using mg/L units.
16. Repeat Steps 3–5. Record the conductivity reading of the desalinated water inside the test tube. _____ mg/L

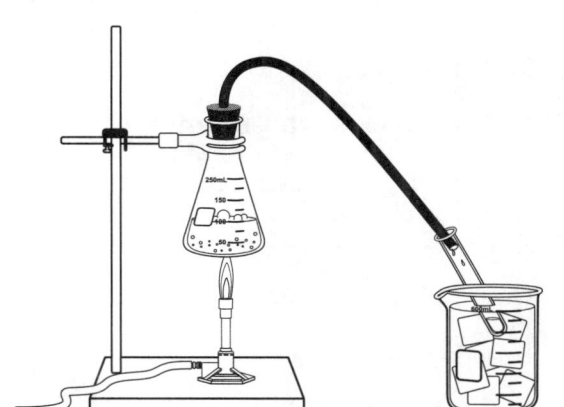

© *Earth and Space Science* • Oceans

Lab 4.2.6A Distillation

ANALYZE AND CONCLUDE:

1. Which sample had a higher conductivity reading? Why? _____
_____
_____
_____
_____
_____

2. What is the most important variable that contributes to an effective separation in a distillation? _____

3. What factors may have contributed to a positive conductivity reading for the desalinated water sample? _____
_____
_____
_____
_____

4. What other methods would be more efficient to produce a desalinated water sample? _____
_____
_____
_____
_____
_____
_____
_____

Name: _____   Date: _____

Surface Currents
WS 4.2.2A

Use Internet resources to locate the following surface currents. Sketch the currents on **WS 4.2.2B Currents Map**, using red for warm currents and blue for cold currents.

Agulhas	Equatorial Countercurrent	Peru
Alaska	Gulf Stream	South Atlantic
Benguela	Kuroshio	South Equatorial
Brazil	Labrador	South Indian
California	North Atlantic	South Pacific
Canary	North Equatorial	West Wind Drift
East Australia	North Pacific	Western Australia
East Greenland	Norwegian	
East Wind Drift	Oyashio	

1. Look at the three major oceans: Pacific, Atlantic, and Indian. What patterns do you see in the currents along the perimeter of each ocean? _____

_____

_____

_____

_____

2. Compare the surface current map to the deep thermohaline current map in the Student Edition. Where do the patterns appear similar? Where are they different?

_____

_____

_____

_____

_____

3. Why do the deep ocean currents seem to join the surface ocean currents at certain places? _____

_____

_____

_____

_____

© Earth and Space Science • Oceans

Name: _____ Date: _____

Currents Map

WS 4.2.2B

Use an Internet resource to locate the ocean currents listed on **WS 4.2.2A Surface Currents**. Sketch the currents on the map below. Use red for warm currents and blue for cold currents.

© Earth and Space Science • Oceans

Name: _____ Date: _____

Tidal Range   WS 4.2.3A

Some ports use tide tables to determine the time and height of high tide and low tide using past observations. Prior to the below data being recorded, individuals attempted to predict the height of the water during a 24-hour period.

Water Levels			
Time	**Water Height (meters)**	**Time**	**Water Height (meters)**
12:00 AM	5.8	12:00 PM	8.7
1:00 AM	6.9	1:00 PM	9.2
1:42 AM	7.2	1:02 PM	9.3
2:00 AM	7.1	2:00 PM	8.6
3:00 AM	6.4	3:00 PM	7.1
4:00 AM	5.2	4:00 PM	4.9
5:00 AM	3.7	5:00 PM	2.4
6:00 AM	2.6	6:00 PM	1.2
7:00 AM	2.2	7:00 PM	-1.0
8:00 AM	2.6	7:54 PM	-1.6
9:00 AM	3.9	8:00 PM	-1.5
10:00 AM	5.6	9:00 PM	-1.0
11:00 AM	7.4	10:00 PM	-0.3
		11:00 PM	2.3

1. Plot the data above on the graph or attach a graph that has been created using a computer software program.

2. Determine the times and water heights that high tides and low tides occurred. _____
_____
_____

3. Calculate the tidal range for the area. _____
_____
_____

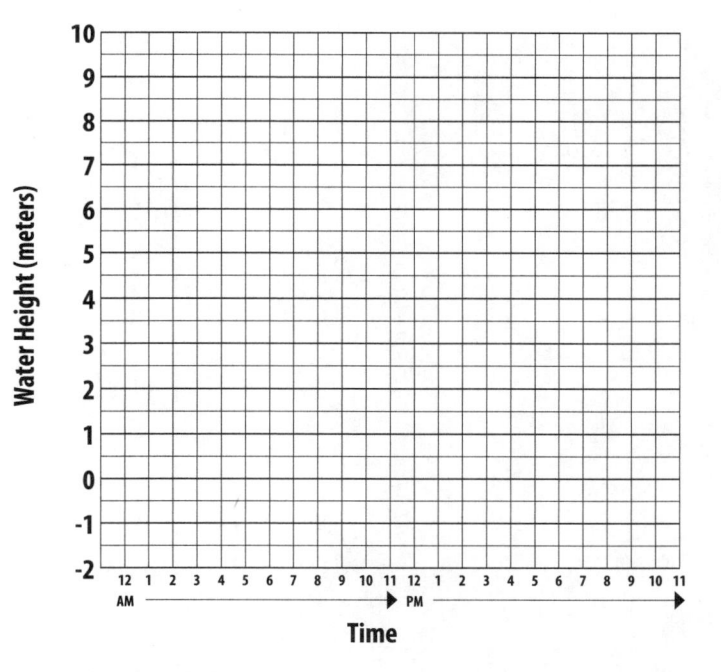

4. Hypothesize when the next high tide will occur. _____
_____
_____

© *Earth and Space Science* • Oceans

Name: _____ **Date:** _____

Oceans vs. Land   WS 4.2.4A

Answer as many questions as you can without discussing them with your classmates.

1. What is the highest mountain on Earth? _____

2. What is the deepest point in Earth's oceans? _____

3. Name the seven continents. _____

4. Name the five major oceans. _____

5. What is the largest island in the world? _____

6. What is the largest sea in the world? _____

7. Who were the first people to land on the moon? _____

8. Who were the first people to climb Mount Everest? _____

9. Who were the first people to dive to the bottom of the Mariana Trench? _____

10. How many people have been on the moon? _____

11. How many people have climbed Mount Everest? _____

12. How many people have been to the bottom of the Mariana Trench? _____

Name: _____ Date: _____

Lab 5.1.1A Percentage of Oxygen

QUESTION: What is the percentage of oxygen in the atmosphere?

HYPOTHESIS: _____
_____

EXPERIMENT:

You will need:	• small candle with stand	• marking pen
• clear, shallow bowl	• matches	
• wide-mouthed glass jar	• food coloring	

Steps:
1. Put about 2 cm of water in the bowl. Invert the jar and place it into the bowl to ensure the water level just reaches over the rim of the jar. Add or remove water as needed. Remove jar and make a small mark on the jar to show the water level.
2. Using centimeters, measure the height of the jar to that mark. _____
3. Add a couple drops of food coloring to the water. Gently stir.
4. Set stand and candle in the center of the bowl. Light the candle.
5. Invert the jar and place it over the candle; leave it until the flame goes out.
6. Mark the side of the jar to show where the new water level is.
7. Measure the distance between the first and second mark. _____
8. Divide the answer in Step 7 by the answer in Step 2. This will give you the percentage of oxygen. _____

ANALYZE AND CONCLUDE:

1. Why did the flame go out? _____

2. What happened to the water level? Why? _____
_____
_____
_____

3. The atmosphere is 21% oxygen. Do your results align with this statement? Why?
_____
_____

4. How do your results compare with other lab groups? Why? _____
_____
_____

© Earth and Space Science • The Atmosphere

Lab 5.1.1A Percentage of Oxygen

5. If you were to do the lab correctly several more times, would you always get the same results? Why? _____

_____

_____

_____

Name: _____ Date: _____

Lab 5.1.1B My Daily Breath

QUESTION: How much of the atmosphere do I breathe each day?

HYPOTHESIS: _____
_____

EXPERIMENT:

You will need:	• plastic tubing	• large basin
• 2 L bottle	• disinfectant wipes	• graduated cylinder

Steps:
1. Breathe normally for 30 seconds, counting your breaths. Multiply by 2 and record your answer. _____
2. Calculate how many breaths you take each day (Answer from Step 1 × 60 × 24). Record your answer. _____
3. Fill the 2 L bottle with water and screw on the cap.
4. Fill the basin half full with water.
5. Invert the bottle and place it in the basin. Take the cap off underwater.
6. Keeping the open end of the bottle underwater, insert one end of the plastic tubing about 3 cm into the bottle.
7. Breathing normally, exhale once through the other end of the tube so the bottle collects the air.
8. Still keeping the opening of the bottle underwater, remove the tubing and replace the cap.
9. Empty the bottle into the graduated cylinder.

    How many mL of water are left? _____
10. Subtract your answer in Step 9 from 2000 mL to determine the volume of air in one breath. _____
11. Calculate the volume of air you breathe in a day. (Step 2 × Step 10) _____

ANALYZE AND CONCLUDE:
1. How much of that air is oxygen? (Step 11 × 21%) _____
2. How much of that air is nitrogen? (Step 11 × 78%) _____
3. Compare your results to other students' results. Why do you think the results are different? _____
_____
_____

© Earth and Space Science • The Atmosphere

Lab 5.1.1B My Daily Breath

4. What errors could have been made during the lab to give incorrect results? _____
_____
_____

5. How much oxygen do you breathe in a week? _____

6. How much nitrogen do you breathe in a week? _____

Name: _____ Date: _____

Lab 5.1.4A Sunscreen

QUESTION: Do higher SPF sunscreens filter out more UV radiation?

HYPOTHESIS: _____
_____

EXPERIMENT:

You will need:	• 4 craft sticks	• 5 sheets of photo-sensitive paper
• 5 clear transparency sheets	• black paper or plastic to cover windows	• tape
• marking pencil		• sunlamp
• sunscreens, SPF 4, 15, 30, 45	• 5 UV meter cards	• stopwatch

Steps:
1. Label one transparency sheet *Control*. Label each of the remaining sheets with a different SPF rating.
2. Using a separate craft stick for each sunscreen, apply thin layer of sunscreen to the appropriate transparency sheet. Spread the sunscreen evenly over the transparency; it does not need to reach the edge of the sheet.
3. Cover the windows with black paper or black plastic and turn out the lights so the room is as dark as possible.
4. Place a UV meter card on a piece of photo paper and cover these with a transparency. If the sunlamp is large enough to expose all 5 sheets at once, set up all 5 sheets. If not, expose 1 sheet at a time and keep the unused photo paper inside the protective package until ready for use.
5. Expose the paper and UV cards to the sunlamp for 5 minutes.
6. Turn off the sunlamp, and using as little light as possible, quickly observe each segment of paper and UV meter card.
7. Record the color of the paper and the UV meter reading in the data table.

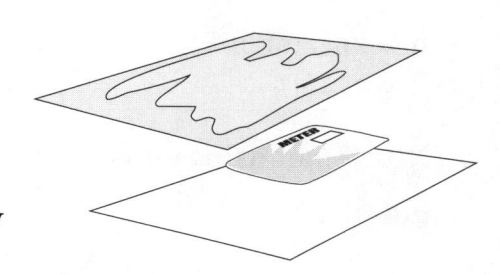

SPF	Color of Paper	UV Meter Reading	Rank
Control			
4			
15			
30			
45			

© Earth and Space Science • The Atmosphere

Lab 5.1.4A Sunscreen

ANALYZE AND CONCLUDE:

1. How does the SPF rating influence the UV protection? _____
_____

2. How might the thinning of the ozone layer influence the sun's effect on human life? _____
_____

3. How do your results compare to the SPF ratings? _____
_____

4. If your results are not what you expected, what might have gone wrong in the experiment? _____
_____

5. Other than using sunscreen, what could you do to protect yourself from UV radiation? _____
_____
_____

6. Do you think the brand of sunscreen or the thickness with which it is applied can affect how well the sunscreen protects from UV radiation? Why? _____
_____
_____

7. Design an experiment that tests different brands of sunscreen or an experiment that tests different thicknesses of the same sunscreen. _____
_____
_____
_____
_____
_____
_____

Name: _____ Date: _____

Air Pollutants
WS 5.1.1A

The following chart contains a list of several air pollutants and their sources. Choose a pollutant to research that may or may not come from the chart. If you choose a different pollutant, add it to the chart.

Air Pollutants

Pollutant	Human Source	Natural Source
Carbon dioxide (CO_2)	Burning fossil fuels	Decay, oceans, forest fires, respiration
Carbon monoxide (CO)	Burning fossil fuels	Forest fires
Sulfur oxides	Burning coal and oil; smelting ores	Volcanoes and decay
Methane (CH_4)	Burning natural gas, decay, termites	Anaerobic bacteria
Nitrogen oxides	Cars	Lightning, bacterial activity
Ammonia (NH_3)	Sewage treatment, decay	Volcanoes, anaerobic bacteria

1. What pollutant did you research? _____

2. In what location is the pollution a problem? _____

3. What types of problems are caused by this pollution? _____
_____
_____

4. What is being done to lower the pollution level? _____
_____
_____

5. What else do you think should be done to lower the pollution level? _____
_____
_____
_____

© Earth and Space Science • The Atmosphere

Name: _____ Date: _____

Layers Chart

WS 5.1.2A

Analyze the layers of the atmosphere.
1. Record whether the layer is part of the homosphere or the heterosphere.
2. Write a description that includes the altitude and temperatures for the layer.
3. Include the jet stream, ozone layer, and ionosphere in the correct layer.

Atmospheric Layers

Layer	Homosphere/Heterosphere	Description
Troposphere		
Stratosphere		
Mesosphere		
Thermosphere		
Exosphere		

© Earth and Space Science • The Atmosphere

Name: _____ Date: _____

Layers
WS 5.1.2B

1. What layer of the atmosphere do you live in? _____

2. In which layer do meteors burn up? _____
_____
_____

3. Which layer in the heterosphere gets warmer as the altitude increases? _____
_____
_____
_____

4. How can the homosphere have such high temperatures but feel so cold? _____
_____
_____
_____
_____

Name: _____ Date: _____

Comparing Theories WS 5.1.3A

Use the chart below to analyze the information given in the text. Summarize what each theory states about each topic. Some topics may not be addressed by both theories.

Comparing Theories

Topic	Dynamo Theory	Rapid-Decay Theory
How did electric current originate?		
A magnetic field exists.		
The magnetic field weakened by 1.4% from 1971–2000.		
The magnetic poles have reversed.		
Mars has a weak magnetic field.		
Mercury does not have a magnetic field.		
Neptune and Uranus have unique magnetic fields.		
What is the age of the earth?		

1. Which topics are facts? _____
_____

2. Which topics are theories? _____
_____

3. What assumptions are made by each theory? _____
_____
_____
_____

4. How do you think a person's worldview will affect a person's opinion about which theory is better? _____
_____
_____
_____

© Earth and Space Science • The Atmosphere

Name: _____ Date: _____

Lab 5.2.1A Heat

QUESTION: Which object will absorb the most heat?

HYPOTHESIS: _____

_____

EXPERIMENT:

You will need:	• sand	• stopwatch
• wide-mouthed jar	• tape	• graph paper
• 4 small thermometers	• small houseplant	• colored pencils
• pie pan	• black construction paper	

Steps:
1. Fill the jar with water and put a thermometer in it.
2. Fill the pie pan with sand and put a thermometer in the middle of the pan. Do not cover the thermometer with sand.
3. Tape a thermometer to the underside of the houseplant's leaves or its stem.
4. Place the black construction paper on the windowsill in direct sunlight and place a thermometer in the center of the paper.
5. Place the jar of water, pan of sand, and houseplant on the windowsill. Be certain that all four objects receive equal amounts of sunlight.
6. Record the temperature on each thermometer every 2 minutes for 10 minutes.

Temperature

Item	2 Minutes	4 Minutes	6 Minutes	8 Minutes	10 Minutes
Water					
Sand					
Houseplant					
Paper					

© Earth and Space Science • Weather

Lab 5.2.1A Heat

7. Make a line graph showing the temperature increase of the four objects. Plot temperature on the *y*-axis and time on the *x*-axis. Include a title for your graph. Use different colored pencils to represent the temperature change of each substance.

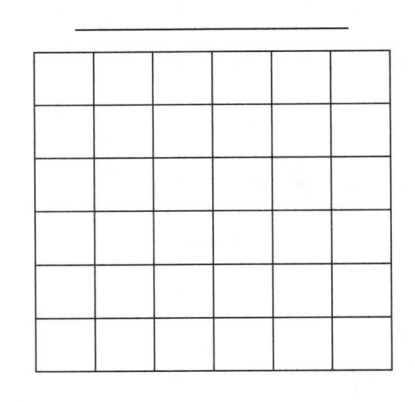

ANALYZE AND CONCLUDE:

1. Which substance absorbed the most heat? Which substance absorbed the least?

_____
_____
_____

2. Which substances heated up the slowest? What areas of the earth might these represent? _____
_____
_____
_____

3. Which substances heated up the fastest? What areas of the earth might these represent? _____
_____
_____
_____

4. Which type of heat transfer caused the items in the lab to heat up? Explain.

_____
_____
_____
_____
_____

© Earth and Space Science • Weather

Name: _____ Date: _____

Lab 5.2.2A Barometric Pressure

QUESTION: How is barometric pressure related to air temperature?

HYPOTHESIS: _____
_____

EXPERIMENT:

You will need:	• beaker	• tape
• glass tube, 50-90 cm in length, closed at one end	• ring stand with clamp	• metric ruler
• colored water	• cardboard strip, 4 cm × 10 cm	
	• marking pencil	

Steps:
1. Fill the glass tube three-fourths full of colored water.
2. Fill the beaker half full of colored water.
3. Cover the open end of the tube with your finger and invert the tube. Lower the tube carefully into the beaker. Be sure the opening of the tube is submerged.
4. Clamp the tube upright on the stand.
5. Tape the scale from **BLM 5.2.2A Barometric Pressure** to the strip of cardboard.
6. Attach the scale to the tube, aligning the zero with the top of the liquid.
7. Record today's date and *0* in the appropriate columns of the data table.

Date	Classroom Barometer Reading	Barometric Pressure	Temperature

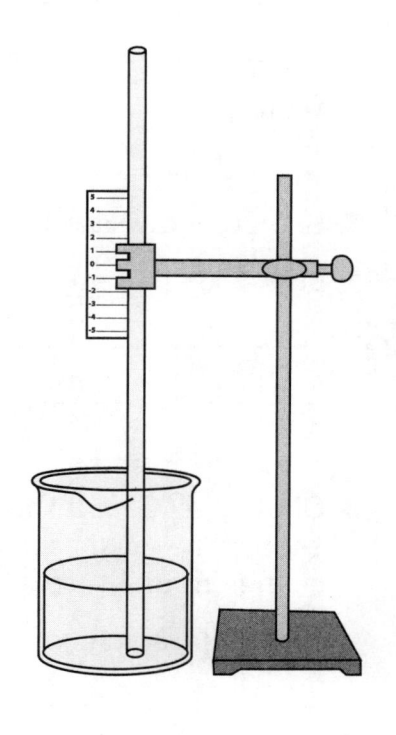

© *Earth and Space Science* • Weather

Lab 5.2.2A Barometric Pressure

8. Research today's temperature and barometric pressure. Record the data in the appropriate location. (Note: If the barometric pressure is given in inches of Hg instead of millimeters, use the following conversion factor: 25.4 × inches of Hg = mm of Hg.)
9. Continue recording data for several days.
10. Make two separate graphs of the data you collected. Use one graph for the barometric pressure and the second graph for the daily temperature. For each graph, place the days on the *x*-axis because this is the independent variable. The barometer and temperature readings should be placed on the *y*-axis because these are dependent variables.

ANALYZE AND CONCLUDE:

1. What day had the highest classroom reading? What day had the highest barometric pressure? Are these on the same day? Why? _____
_____
_____

2. What day had the highest temperature reading? What was it? _____
_____

3. What day had the lowest class reading? What day had the lowest barometric pressure? Are these on the same day? Why? _____
_____
_____

4. Which day had the lowest temperature? What was it? _____
_____

5. Is there a direct relationship or an inverse relationship between barometric pressure and temperature? Why? _____
_____
_____

6. Challenge: Graph the classroom readings. Place the days of the week along the *x*-axis. The *y*-axis should be numbered from −10 mm through 10 mm. Compare the change in pressure in the classroom to the actual barometer changes. Are the lines on the classroom graph and the barometer graph the same? Why? _____
_____

Name: _____ Date: _____

Lab 5.2.3A Anemometer

QUESTION: Does greater wind speed indicate a change in weather?

HYPOTHESIS: _____
_____

EXPERIMENT:

You will need:	• 2 strips of stiff cardboard that are equal in length	• pencil with eraser
• 4 small paper cups	• stapler	• metric ruler
• scissors	• pushpins	• stopwatch
• marker		

Steps:
1. Cut off the rolled edges of the paper cups.
2. Color the outside of one cup with the marker.
3. Cross the cardboard strips to make a plus sign and staple them together.
4. Staple the cups to the ends of the cardboard strips so each cup opening is at a right angle to the next one.
5. Use the ruler to find the exact center of the intersection of the cardboard strips. Push the pushpin through the center of the cardboard strips and into the eraser of the pencil.
6. Take the anemometer outside. Holding the anemometer straight up, count the number of times the colored cup spins around in 1 minute. (Note: You are measuring the wind speed in turns or revolutions per minute. This is not an accurate form of measurement, so take measurements three times and then find the average.) Record your results. Do this for 7 days.
7. Calculate the number of revolutions per hour. Record your results.
8. Find the actual wind speed for your area on the Internet and record it. Weather stations record wind speed in kph or mph. Your numbers will not be the same as actual wind speed, but you should observe an increase or decrease in wind speed similar to the increase or decrease in actual wind speed.

Date	Trial 1	Trial 2	Trial 3	Average	Rotations per Hour	Actual Wind Speed

© Earth and Space Science • Weather

Lab 5.2.3A Anemometer

ANALYZE AND CONCLUDE:

1. On which day did you have the highest number of revolutions per hour? Was it on the same day as the highest reading for the actual wind speed? _____

2. On which day did you have the lowest number of revolutions per hour? Was it on the same day as the lowest reading for the actual wind speed? _____

3. What problems may have affected the accuracy of your anemometer? _____

4. What improvements could you make to get more accurate results? _____

5. On the days with higher wind speed, did you see a change in weather? If yes, what was the change? _____

6. What causes wind and why do meteorologists measure it? _____

Name: _____ Date: _____

Lab 5.2.5A Cloud Formation

QUESTION: Will a cloud form?

HYPOTHESIS: _____
_____

EXPERIMENT:

You will need:	• isopropyl alcohol	• duct tape
• safety goggles	• 2 L bottle	• bicycle pump

Steps:
1. Put on safety goggles.
2. Pour 2 tbsp of alcohol into the bottle.
3. Screw the lid on the bottle, hold it horizontally, and then slowly rotate it. Some of the alcohol will evaporate. At this point, the bottle contains liquid alcohol, free alcohol molecules, and water vapor.
4. Remove the lid and place a piece of duct tape over the opening of the bottle, creating a tight seal.
5. Poke a small hole in the tape over the opening.
6. Pressurize the bottle by pumping air through the hole in the duct tape. Hold the air pump tightly against the duct tape, sealing the opening. Continue pumping until you cannot pump any more air in the bottle. This increases air pressure and temperature.
7. Keeping your face away from the opening of the bottle, pull the pump away from the bottle. What happened? _____

ANALYZE AND CONCLUDE:

1. Did a cloud form? _____

2. How does a cloud form in this system? _____
_____
_____

3. If a cloud did not form, what could be the problem with the system? _____
_____
_____
_____

4. How is the lab activity similar to real clouds forming? _____
_____
_____

© Earth and Space Science • Weather

Name: _____ Date: _____

Lab 5.2.6A Relative Humidity

QUESTION: How does temperature affect relative humidity?

HYPOTHESIS: _____

_____

EXPERIMENT:

You will need:	• ice water	• rubber band
• thermometer	• room-temperature water	• heat-resistant gloves
• hot plate	• three 7 cm × 7 cm squares of cotton fabric	• small piece of cardboard
• 3 beakers		

Steps:
1. Place the thermometer on a flat surface for 5 minutes. Record the temperature. This temperature reading is the dry bulb temperature for all three trials.
2. Pour ice water into one beaker and room-temperature water into another beaker.
3. Heat some water in the third beaker over the hot plate; do not exceed 34°C.
4. Secure a square of cloth around the bulb of the thermometer with the rubber band, and moisten the cloth in the hot water.
5. Wearing heat-resistant gloves, remove the thermometer from water. Hold the top of the thermometer and rapidly fan the thermometer bulb with the cardboard for 1 minute.
6. Quickly unwrap the cloth and record the wet bulb temperature.
7. Subtract the temperature of the wet bulb from the temperature of the dry bulb and record the difference. _____
8. Use the data table to determine the relative humidity of the cloth.
9. Repeat Steps 4–8 with the beakers of ice water and room-temperature water.

Water	Water Temperature	Dry Bulb Temperature	Wet Bulb Temperature	Temperature Difference	Relative Humidity
Hot					
Ice water					
Room Temperature					

ANALYZE AND CONCLUDE:

1. Which water temperature had the lowest relative humidity and which had the highest? _____

_____

_____

© Earth and Space Science • Weather

Lab 5.2.6A Relative Humidity

2. Why did the temperatures produce different relative humidities? _____

_____

_____

_____

_____

_____

3. Why might you feel uncomfortable on a hot, humid day? _____

_____

_____

4. If your results are not what you expected, explain why you think that is.

_____

_____

Relative Humidity (%)

Dry Bulb Temp °C	\multicolumn{10}{c}{Difference Between Dry Bulb and Wet Bulb Temperature °C}
	1	2	3	4	5	6	7	8	9	10
0	81	64	46	29	13					
2	84	68	52	37	22	7				
4	85	71	57	43	29	16				
6	86	73	60	48	35	24	11			
8	87	75	63	51	40	29	19	8		
10	88	77	66	55	44	34	24	15	6	
12	89	78	68	58	48	39	29	21	12	
14	90	79	70	60	51	42	34	26	18	10
16	90	81	71	63	54	46	38	30	23	15
18	91	82	73	65	57	49	41	34	27	20
20	91	83	74	66	59	51	44	38	31	24
22	92	83	76	68	61	54	47	41	34	28
24	92	84	77	69	62	56	49	44	37	31
26	92	85	78	71	64	58	51	47	40	34
28	93	85	78	72	65	59	53	48	42	37
30	93	86	79	73	67	61	55	50	44	39
32	93	87	80	74	68	63	57	52	47	42
34	93	87	81	75	69	64	59	54	49	44

Name: _____ Date: _____

Lab 5.2.7A Air Mass Interaction

QUESTION: How do cold and warm air masses interact?

HYPOTHESIS: _____

_____

EXPERIMENT:

You will need:	• red and blue food coloring	• plastic card, slightly larger than
• 2 identical, wide-mouthed jars	• spoon	mouth of the jars
• hot and cold water	• large pan	

Steps:
1. Fill one jar with cold water and the other jar with hot water. Fill the jars to the brim, almost to the point of spilling over.
2. Add red food coloring to the hot water. Add blue food coloring to the cold water. Gently stir each jar of water with the spoon.
3. Place the jar with the cold water in the pan. Put the plastic card over the jar opening. Holding the card over the opening, flip the jar upside down. A vacuum should be created and the card should stay in place on the mouth of the jar.
4. Place the warm water jar in the pan.
5. Carefully set the inverted cold water jar on top of the warm water jar. Check all around the jars to be sure the openings of the jars are perfectly aligned.
6. Slowly remove the card from between the two jars. Watch the water in the jars and note what happens.
7. Empty the jars in the sink.
8. Repeat Steps 1 and 2.
9. Place the jar with the hot water in the pan. Put the plastic card over the jar opening. Holding the card over the opening, flip the jar upside down.
10. Place the cold water jar in the pan.
11. Carefully set the hot water jar on top of the cold water jar. Check all around the jars to be sure the openings of the jars are perfectly aligned.
12. Slowly remove the card from between the two jars. Note what happens.

ANALYZE AND CONCLUDE:

1. What happened in Step 6? Why? _____

_____

_____

© Earth and Space Science • Weather

Lab 5.2.7A Air Mass Interaction

2. What happened in Step 12? Why? _____
_____
_____
_____

3. How is this experiment like a warm air mass and a cold air mass coming in contact with each other? _____
_____
_____
_____
_____

4. What problems did you encounter during the lab? _____
_____
_____
_____
_____
_____
_____

5. If your results were not what you expected, what might have gone wrong?
_____
_____
_____
_____
_____
_____

Name: _____ Date: _____

Daily Temperature   WS 5.2.1A

Temperatures change daily. Convection currents in the air and cloud cover both contribute to these changes. Record the outdoor temperature and cloud cover each day for five days. Do this at the same time each day.

Daily Temperature and Cloud Cover

Day and Date	Outside Temperature	Amount of Cloud Cover

1. Which day had the highest temperature? How much cloud cover was there? _____
_____
_____

2. Which day had the lowest temperature? How much cloud cover was there? _____
_____
_____

3. From your observations, what can affect daily temperatures? Why? _____
_____
_____
_____
_____
_____
_____

© Earth and Space Science • Weather

Name: _____   Date: _____

Global Winds
WS 5.2.4A

Label the *equator*, *30° latitude lines*, *60°N latitude line*, and the *North and South Poles*. Label the following winds and draw arrows to indicate the direction each wind flows: *trade winds*, *polar easterlies*, and *westerlies*.

1. What factors create different global wind patterns? _____
_____
_____
_____
_____

2. Which global wind patterns affect your location? _____
_____
_____
_____
_____

© Earth and Space Science • Weather

Name: _____   Date: _____

Cloud Observations

WS 5.2.5A

Mark all boxes that correspond to your observations. Record daily temperature and precipitation.

Date	Time	Cloud types (more than one may apply)	Cirrus	Cirrocumulus	Cirrostratus	Altocumulus	Altostratus	Stratus	Stratocumulus	Nimbostratus	Cumulus	Cumulonimbus	Cloud Cover	Clear (<10% cover)	Scattered (10%–50% cover)	Broken (50%–90% cover)	Overcast (90%–100% cover)	Altitude	Low	Middle	High	Temperature	Precipitation

© Earth and Space Science • Weather

Name: _____   Date: _____

Average Precipitation   WS 5.2.6A

Using the given data, make a bar graph showing the average rainfall and snowfall for Bismarck, North Dakota. The averages were taken from data collected between 1981 and 2010.

Average Precipitation in Bismarck, North Dakota

Month	Average Rainfall in cm	Average Snowfall in cm
January	1.1	22.9
February	1.3	20.3
March	2.2	22.9
April	3.2	10.2
May	6.1	0
June	8.1	0
July	7.3	0
August	5.8	0
September	4.0	0
October	3.2	5.1
November	1.8	22.9
December	1.2	22.9

1. Place the months of the year along the x-axis because these are the independent variables.
2. Along the y-axis, place the average precipitation. Mark in increments of 1 cm, leaving an extra line between each centimeter for ease in marking decimals. The range will be from 0 to 23 cm.
3. Use one color to graph the rainfall and another color to graph the snowfall.
4. Which months had more rain than snow? _____
5. Which month had the most rain? _____
6. Which month had the most snow? _____
7. What is the total average rainfall? What is the total average snowfall? _____

8. Does Bismarck receive more rain or snow each year? Why do you think that is?

_____
_____
_____

© Earth and Space Science • Weather

Name: _____ Date: _____

Air Masses   WS 5.2.7A

Draw two of each type of air mass on the map below. Use the abbreviations in the Student Edition to label each air mass. Color continental polar air masses blue, continental tropical air masses yellow, maritime polar air masses green, maritime tropical air masses orange, and arctic air masses purple.

1. If a maritime tropical air mass is in the area, what weather would you expect to have? _____

2. What change in weather in an area would occur if the maritime tropical air mass was followed by a continental tropical air mass? _____
_____

3. Would a continental polar or a maritime polar air mass be more likely to bring snow to an area? _____ Why? _____
_____

4. If a continental polar air mass was followed by an arctic air mass, what change in the weather would occur? _____
_____

Why? _____
_____

© Earth and Space Science • Weather

Name: _____ Date: _____

Stormy Weather

WS 5.2.8A

Illustrate the stormy weather.

1. Draw the three stages of a thunderstorm. Include names of the stages, changes in height of the clouds, arrows indicating updrafts and downdrafts, and types of precipitation.

2. Draw the three steps of cloud-to-ground lightning formation. Include the cloud in each step. Show where the charges are in the cloud and on the ground. Include a tall object on the ground where the positive charges can accumulate. Indicate the parts of the lightning.

3. In which stage of a thunderstorm does the most precipitation fall? What types of precipitation can fall? _____

_____

4. What type of cloud produces lightning? _____

5. What causes lightning to form? _____

_____

_____

_____

6. Why is it dangerous to shelter underneath a tree during a thunderstorm? _____

_____

_____

© Earth and Space Science • Weather

Name: _____   Date: _____

On-the-Scene Report

WS 5.2.8B

Research information about a specific tornado or hurricane. Answer the following questions to gather information to use for the on-the-scene report:

1. Type of storm: _____
_____

2. What damage was done by the storm? _____
_____

3. Where did the storm take place? _____

4. When did it happen? Date? Time of day? Season? _____
_____
_____

5. How did the storm form? How did it affect human life, businesses, homes, and schools? _____
_____
_____
_____

6. How did first responders help people affected by the storm? _____
_____
_____
_____

7. Why would people want to know about the storm? _____
_____
_____

8. Were there any more storms after the initial storm? If so, give details. ____
_____
_____

9. List any other information that will be useful for presentation. _____
_____
_____
_____
_____

© Earth and Space Science • Weather

Name: _____ Date: _____

Weather Predictions   WS 5.2.9A

Use the information gathered from **WS 5.2.5A Cloud Observations** and weather history from the Internet to complete the following chart:

Daily Weather

Date	High Temp	Low Temp	High Pressure	Low Pressure	Precipitation	Cloud Types

1. Make a line graph showing the high and low temperature for each day. Use red to connect high temperatures. Use blue to connect low temperatures. Label the *x*-axis *Days*; label the *y*-axis *Temperature*.

Weather Predictions

WS 5.2.9A

2. Make a graph showing high and low pressure for each day. Use red to connect high pressures. Use blue to connect low pressures. Label the *x*-axis *Days*; label the *y*-axis *Pressure*.

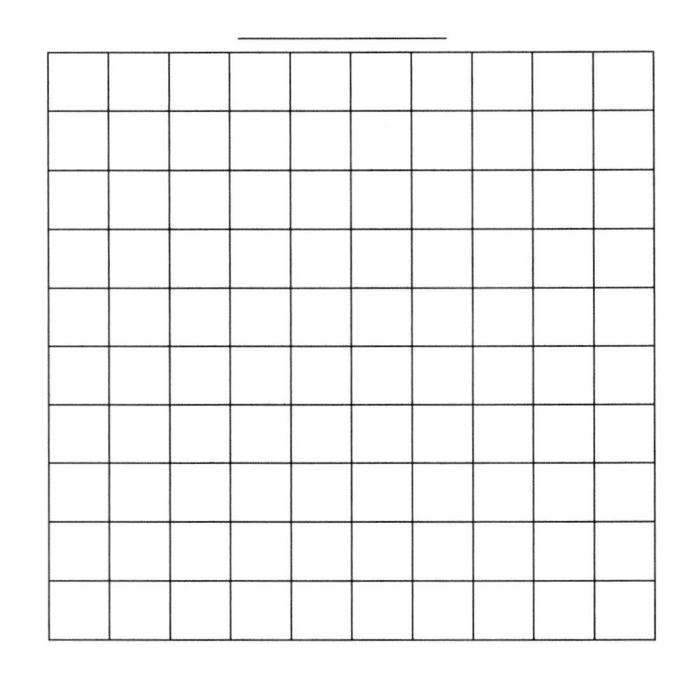

3. Compare the graph of the temperatures to the graph of the pressures. Do you see a direct relationship or an inverse relationship? Explain. _____
_____

4. Look at the data and graphs and make a prediction for tomorrow's weather. Include all data listed in the table. _____
_____

5. What was the actual weather for that day? _____
_____
_____

6. How does your prediction compare with the actual weather? What is the same, and what is different? _____
_____

7. If your prediction was not correct, explain why. _____
_____
_____

© Earth and Space Science • Weather

Name: _____ Date: _____

Weather Map

WS 5.2.9B

Using the weather map provided, answer the following questions:

1. How many high-pressure systems are shown on the weather map? Where are the centers located? _____
_____

2. How many low-pressure systems are shown on the map? Where are the centers located? _____
_____

3. What other kinds of fronts are on the map and where are they located? _____
_____
_____

4. Where is the weather likely to be clear and dry? _____
_____

5. Is it likely to be stormy anywhere? _____
_____
_____

6. Where is the highest temperature on the map? _____

7. What is the lowest temperature on the map? _____

8. Describe the weather in your area currently. _____
_____
_____

9. What type of weather is forecasted for your area? _____
_____
_____
_____

© Earth and Space Science • Weather

Name: _____ Date: _____

Weather Report Evaluations

WS 5.2.9C

Evaluate the weather report of each group. Evaluate your own group as well.

1. Group name: _____

2. Severe weather type: _____

3. Description of weather event given in report: _____
_____
_____
_____

4. What did the group do well? _____
_____
_____
_____

5. What needs improvement? _____
_____
_____
_____

1. Group name: _____

2. Severe weather type: _____

3. Description of weather event given in report: _____
_____
_____
_____

4. What did the group do well? _____
_____
_____
_____

5. What needs improvement? _____
_____
_____
_____

© Earth and Space Science • Weather

Weather Report Evaluations

WS 5.2.9C

Evaluate the weather report of each group. Evaluate your own group as well.

1. Group name: _____

2. Severe weather type: _____

3. Description of weather event given in report: _____
_____
_____
_____

4. What did the group do well? _____
_____
_____
_____

5. What needs improvement? _____
_____
_____
_____

1. Group name: _____

2. Severe weather type: _____

3. Description of weather event given in report: _____
_____
_____
_____

4. What did the group do well? _____
_____
_____
_____

5. What needs improvement? _____
_____
_____
_____

© Earth and Space Science • Weather

Name: _____ Date: _____

Lab 5.3.1A Earth's Axial Tilt

QUESTION: How does latitude determine climate?

HYPOTHESIS: _____
_____

EXPERIMENT:

You will need:	• flashlight	• protractor
• 4 large sheets of paper	• metric ruler	

Steps:
1. Tape one large sheet of paper to a flat surface.
2. Shine the flashlight in a vertical position, 20 cm above the piece of paper.
3. Trace the outline of the ellipse of light made by the flashlight. Label the sheet of paper *90 Degrees*.
4. Repeat Steps 1 through 3 with the flashlight positioned at a 70°, 45°, and 20° angle. Make sure that the flashlight remains 20 cm above the pieces of paper.

ANALYZE AND CONCLUDE:

1. What angle cast the smallest beam of light? _____

2. What angle cast the largest area of light? _____

3. Predict which angle would produce the most heat. Explain your prediction.

_____
_____
_____
_____

4. How is this activity related to climate? _____

_____
_____
_____
_____

5. Why is the latitude of a region not a complete indicator of climate patterns?

_____
_____

© Earth and Space Science • Climate

Name: _____ Date: _____

Lab 5.3.2A Microclimates

QUESTION: What determines a microclimate?

HYPOTHESIS: _____
_____

EXPERIMENT:

You will need:	• metric ruler	• soil color chart
• thermometer	• trowel	
• hand lens	• paper towels	

Steps:
1. Select a small location on the school grounds that has unique climate conditions compared to the surrounding area. Draw a map of the school grounds and mark the particular location you will be examining.

2. Provide details regarding the weather conditions on the day of experimentation. Is there any cloud cover, strong winds, or high humidity levels? _____
_____
_____

3. Using a thermometer, measure the air temperature in the small location. Make sure the thermometer is not in direct sunlight as this will give an inaccurate air temperature measurement. After five minutes has elapsed, record the thermometer reading.

   Air temperature: _____ °C

4. Apply a dry paper towel to the soil surface. Classify the soil surface as extremely moist, somewhat moist, dry, or extremely dry. _____

© Earth and Space Science • Climate

Lab 5.3.2A Microclimates

5. Dig a small 15 cm hole and apply another dry paper towel to the soil below ground. Classify the soil found below ground as extremely moist, somewhat moist, dry, or extremely dry. _____

6. Note the color of the soil using the soil color chart, if available. _____
_____

7. Describe the vegetation in the small location. _____
_____

8. Measure and record the height of the tallest, unaltered plant. If the tallest, unaltered plant is a tree, estimate the height of the tree in meters.

   Plant height: _____ cm

9. Repeat Steps 2 through 6 over a period of four days and record data in the table.

	Weather Conditions	Air Temperature (°C)	Soil Surface Moisture Content	Below Ground Soil Moisture Content	Soil Color
Day 2					
Day 3					
Day 4					
Day 5					

ANALYZE AND CONCLUDE:

1. Research the daily temperatures and precipitation values for the surrounding climate and record in the table.

	Air Temperature (°C)	Amount of Precipitation (cm)
Day 2		
Day 3		
Day 4		
Day 5		

© Earth and Space Science • Climate

Name: _____ Date: _____

Lab 5.3.2A Microclimates continued

2. How does the microclimate compare to the surrounding overall climate? _____
_____

3. Would the observed conditions be the same during a different season? Explain.
_____
_____
_____

4. If anomalies are observed in the microclimate, what is responsible for the unusual conditions? _____
_____

5. Compare air temperature values with other groups and graph the results.

Temperature

Location 1   Location 2   Location 3   Location 4

© Earth and Space Science • Climate

Name: _____ Date: _____

Lab 5.3.3A Greenhouse Gases

QUESTION: How might excess greenhouse gases warm the earth?

HYPOTHESIS: _____

_____

EXPERIMENT:

| **You will need:** • 3 thermometers | • 2 glass jars • paper towels | • sun lamp |

Steps:
1. Place three thermometers in direct sunlight on a windowsill or under a sun lamp until all three thermometers read the same temperature. Thermometers should be positioned close to one another, at least five centimeters apart. Record the temperature. _____ °C
2. Position the mouths of two upside-down glass jars over the bulbs of two of the thermometers. The third thermometer should remain uncovered.
3. Place a moistened paper towel under one of the glass jars. Do not allow the paper towel to come in contact with the thermometer.
4. Observe and record the temperature of all three thermometers every two minutes in the table below.

Time (min)	Temperature (°C)		
0	**Thermometer 1**	**Thermometer 2**	**Thermometer 3**
2			
4			
6			
8			
10			
12			
14			
16			
18			
20			
22			
24			
26			
28			
30			
32			

© *Earth and Space Science* • Climate

Lab 5.3.3A Greenhouse Gases

ANALYZE AND CONCLUDE:

1. Why is it important that all three thermometers be the same temperature at the beginning of the laboratory experiment? _____
_____
_____

2. What greenhouse gas was tested in this experiment? _____
_____

3. Graph the three sets of temperature readings over time in the area below.

Dry, Covered Thermometer	Moist, Covered Thermometer	Uncovered Thermometer
Temperature (°C) vs Time (min)	Temperature (°C) vs Time (min)	Temperature (°C) vs Time (min)

4. How do the three sets of temperature readings compare? _____
_____
_____
_____

5. Explain why the thermometer under the jar with the moistened thermometer had the greatest temperature increase over time. _____
_____
_____
_____

Name: _____ Date: _____

Rain Shadow Effect

WS 5.3.1A

1. Draw blowing winds with arrows over the mountain image below. Cool, moist air should be indicated by blue arrows flowing over the windward side of the mountain. Dry, hot air should be indicated by red arrows flowing over the leeward side of the mountain. Label the windward and leeward sides of the mountain. On the appropriate sides, draw the growth of vegetation, clouds producing rain or snow, and barren land.

2. Choose a city on the windward side of a mountain and a city on the leeward side of that same mountain. Research the annual precipitation and annual high and low temperature of each city. Using an Internet resource, determine the precipitation and temperatures of each city for one week.

	Windward City _____	Leeward City _____
Annual precipitation (cm)		
Average high temperature (°C)		
Average low temperature (°C)		
Precipitation (cm), Day 1		
Precipitation (cm), Day 2		
Precipitation (cm), Day 3		
Precipitation (cm), Day 4		
Precipitation (cm), Day 5		
Precipitation (cm), Day 6		
Precipitation (cm), Day 7		

Earth and Space Science • Climate

Rain Shadow Effect

WS 5.3.1A

	Windward City _____	Leeward City _____
High Temperature (°C), Day 1		
Low Temperature (°C), Day 1		
High Temperature (°C), Day 2		
Low Temperature (°C), Day 2		
High Temperature (°C), Day 3		
Low Temperature (°C), Day 3		
High Temperature (°C), Day 4		
Low Temperature (°C), Day 4		
High Temperature (°C), Day 5		
Low Temperature (°C), Day 5		
High Temperature (°C), Day 6		
Low Temperature (°C), Day 6		
High Temperature (°C), Day 7		
Low Temperature (°C), Day 7		

3. How do the leeward city temperatures and precipitation compare to the windward city temperatures and precipitation? _____

_____

_____

_____

_____

© Earth and Space Science • Climate

Name: _____ **Date:** _____

La Paz Climate Analysis   WS 5.3.1B

The purpose of this activity is to observe patterns in data regarding precipitation, temperature, elevation, and any land features such as oceans, lakes, and mountains for the city of La Paz, Bolivia. These patterns help explain local climate conditions.

1. Provide the following information regarding the location of La Paz, Bolivia:

- Latitude: _____
- Longitude: _____
- Elevation (meters): _____

2. Is the location near large bodies of water? _____

3. Observe prevailing winds and describe whether the wind blows across water or land to reach La Paz. Note any seasonal changes. _____
_____
_____
_____
_____

4. Is La Paz near mountains? _____

5. Do prevailing winds cross the mountains or La Paz first? Note any seasonal changes. _____
_____

6. Research and record the average temperature (°C) and precipitation (mm) for each month.

Average Temperature (°C)												
Jan	Feb	Mar	Apr	May	Jun	Jul	Aug	Sep	Oct	Nov	Dec
Average Precipitation (mm)												

© *Earth and Space Science* • Climate

La Paz Climate Analysis

WS 5.3.1B

7. Create a Combo Clustered Column-Line Climograph below using the above collected data. Indicate months on the *x*-axis. Include temperature measurements on the primary (left) *y*-axis. Represent this data in the form of a bar graph. Indicate precipitation on the secondary (right) *y*-axis. Represent this data in the form of a line graph.

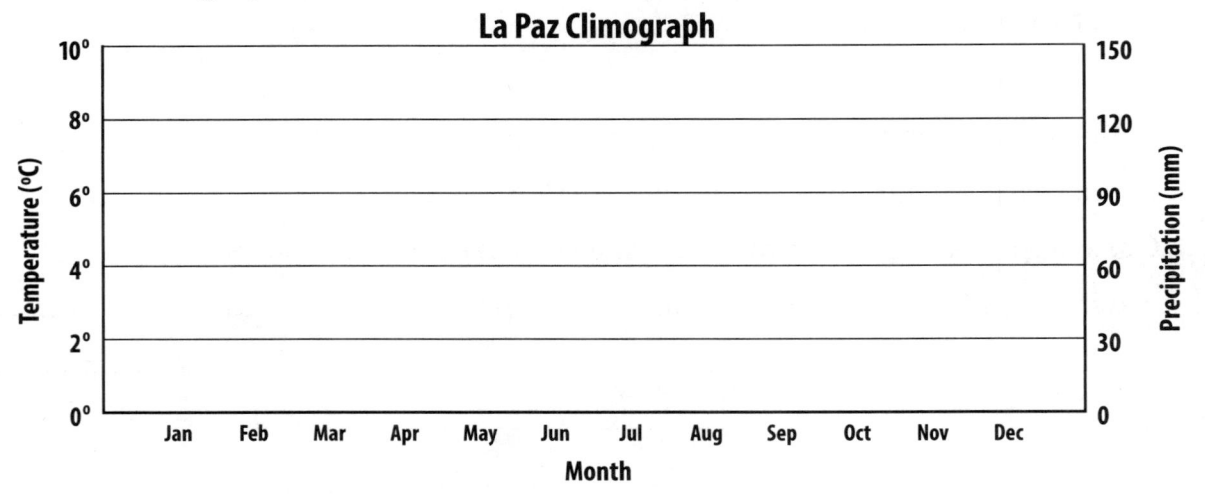

La Paz Climograph

[Blank climograph grid with Temperature (°C) on left y-axis ranging 0°–10°, Precipitation (mm) on right y-axis ranging 0–150, and Months (Jan–Dec) on x-axis]

8. Calculate the average precipitation for one year. Is there a pattern? _____

9. Calculate the average temperature for one year. _____

10. Calculate the range of temperature for one year. Note any seasonal changes. _____

11. Calculate the range of precipitation for one year. Note any seasonal changes. _____

12. What are the hottest months? _____

13. What are the coldest months? _____

14. Write a summary about how proximity to large bodies of water, elevation, and mountains affects temperature and precipitation. _____

© Earth and Space Science • Climate

Name: _____   Date: _____

Local Climate
WS 5.3.2A

1. Locate your city or town on a world map. Note the latitude. _____

2. Using an Internet resource, determine the following information: _____

   • Annual rainfall: _____ cm

   • Average temperature: _____ °C

   • Annual temperature range: _____ cm

3. What type of seasons occur in the local area? _____
   _____

4. Describe the predominant vegetation found in the local area. _____
   _____

5. Provide details regarding the local topography. Include mountain ranges and large bodies of water. _____
   _____
   _____

6. From the gathered information, determine the local climate region. _____
   _____

7. Why did you choose this particular type of climate region? _____
   _____

8. Is the climate the same as it was 50, 100, and 1,000 years ago? _____
   _____

9. How was the lifestyle of the native people in the local area affected by the climate? _____
   _____

10. How is the lifestyle of people currently affected by the climate? _____
    _____
    _____

11. How would the lifestyle or economy change if the climate suddenly changed?
    _____
    _____
    _____

© Earth and Space Science • Climate

Name: _____ Date: _____

My Contribution
WS 5.3.3A

Perform the following calculations to estimate how much carbon dioxide is released into the atmosphere each month from the vehicle that transports you to and from school.

1. Determine the distance in kilometers from home to school.

2. Calculate how many kilometers you travel to and from school each month.

3. Locate the estimated kilometers per liter value provided for the vehicle that transports you to and from school. Calculate the total number of liters of gasoline used during your trips to and from school using the distance derived in Step 2. If the estimated value provided is miles per gallon, you will need to convert to kilometers per liter. (1 mile = 1.61 km; 1 gallon = 3.79 L)

4. Burning one liter of gasoline produces about 2.5 kg of carbon dioxide. How much carbon dioxide is released? _____

5. What could you do that could not only help you be a good steward of natural resources but also possibly reduce your contribution of carbon dioxide? _____
_____
_____
_____
_____

© Earth and Space Science • Climate

Name: _____ Date: _____

Lab 6.1.2A Plants and the Atmosphere

QUESTION: How do plants help renew the atmosphere?

HYPOTHESIS: _____

_____

EXPERIMENT:

You will need:	• petroleum jelly	• thermometer
• 2 paintbrushes	• plastic wrap	• soil moisture sensor
• 2 identical small plants	• 2 clear, plastic bags	

Steps:
1. Have two group members use the paintbrushes to coat both sides of all the leaves on one of the plants with petroleum jelly.
2. Saturate the soil of both plants with equal amounts of water.
3. Cover the soil of both plants with plastic wrap and place a plastic bag over each plant.
4. Place both plants on a windowsill in direct sunlight and observe for several days.
5. After several days have elapsed, measure the temperature of both soils.

   • Non-petroleum plant soil: _____ °C

   • Petroleum plant soil: _____ °C

6. Predict which plant will have the greater soil moisture content. _____
   _____

7. Use the soil moisture sensor to measure the water content percentage in both soils.

   • Non-petroleum plant soil: _____ %

   • Petroleum plant soil: _____ %

ANALYZE AND CONCLUDE:

1. Describe the appearance of the petroleum-covered plant after several days have passed. Explain why the plant appears this way. _____

_____
_____
_____
_____
_____

Lab 6.1.2A Plants and the Atmosphere

2. Which plastic bag held the most water? Where did the water come from? _____

3. Correlate the soil water contents to the amount of water produced on the inside of the plastic bags. _____

4. Explain why the plastic bag around the plant that was not covered with petroleum appears inflated. _____

5. How do green plants help renew the atmosphere? _____

Earth and Space Science • Natural Resources

Name: _____ Date: _____

Lab 6.1.3A Oil Reserve Model

QUESTION: How do individuals locate oil reserves?

HYPOTHESIS: _____
_____

EXPERIMENT:

You will need:	• balloon	• graph paper
• sand	• syringe	• metric ruler
• small rocks	• viscous liquid	• probe
• cardboard box		

Steps:
1. Place sand and small rocks in a cardboard box such as a shoebox.
2. Fill a balloon with the viscous liquid using a syringe, and place the balloon in a layer of sand and rock.
3. Attach graph paper to the lid of the box, and poke holes large enough to insert a probe at each gridline cross section. Mark an *X* on one side of the lid, and secure the lid on the box.
4. Map the location of the "oil reserve" and exchange boxes with another group of students.
5. Do not adjust or open the lid of the other students' box.
6. Tap the lid of the box to estimate by sound where the oil reserve is located. Mark the areas on the graph paper that you think should be probed.
7. Use a ruler and mark 1 cm increments, beginning at the bottom, on the probe.
8. Probe gently into the areas you designated to be probed until the oil reserve has been tapped.
9. Monitor how many times the probe was inserted into the box and to what depth. Each centimeter costs $200,000.00, and each time the probe is removed and reinserted, a cost of $100,000.00 is applied.

ANALYZE AND CONCLUDE:

1. Calculate how much your oil tapping adventure cost. _____

2. What changes would you make to the procedure? _____
_____
_____

3. How did the oil reserve's estimated location compare to the actual location determined by the group that created the box? _____
_____
_____

© *Earth and Space Science* • Natural Resources

Name: _____ Date: _____

Lab 6.1.4A Mining Minerals

QUESTION: How are minerals extracted from the ground?

HYPOTHESIS: _____
_____

EXPERIMENT:

You will need:	• 2 gold beads or sequins	• 16 green beads or sequins
• birdseed mix	• 4 silver beads or sequins	
• shallow pan	• 8 red beads or sequins	

Steps:
1. Pour the birdseed in the shallow pan.
2. Add the beads to the pan, and stir the mixture well.
3. Spend five minutes searching through the mixture and "mining" the various seeds and beads. The sunflower seeds represent iron, the other seeds represent waste, the gold beads represent gold, the silver beads represent silver, the red beads represent copper, and the green beads represent zinc. Place the mined products on a piece of paper in separate piles.
4. At the end of five minutes, count each kind of product from the piles made, and record the quantity in the table below.

	Gold	Silver	Copper	Zinc	Iron	Waste
Quantity						

5. Count and record the remaining millet grains scattered in the pan. _____

ANALYZE AND CONCLUDE:

1. Multiply the quantity of each mineral by the designated value of each mineral in the table below.

Mineral	Quantity	Value	Total for each
Gold		$8.00	
Silver		$6.00	
Zinc		$3.00	
Copper		$2.00	
Iron		$1.00	

2. Calculate the total for all minerals. _____

3. Multiply the quantity of millet grains by $5.00. This represents the cleanup cost. Subtract this amount from the total for all minerals. _____

© *Earth and Space Science* • Natural Resources

Lab 6.1.4A Mining Minerals

4. If any additional millet grains remained in the pan, multiply this quantity by $5.00. This represents the environmental damage penalty. _____

5. What is your total profit? (Total for all minerals – cleanup cost – environmental damage penalty) _____

6. Was mining for gold or iron more profitable? Why? _____
_____
_____

7. What is the most profitable, responsible way for miners to mine? Why? _____
_____
_____
_____
_____
_____
_____
_____

Name: _____ Date: _____

Lab 6.1.4B Mining Conditions

QUESTION: How does mining affect the land?

HYPOTHESIS: _____
_____

EXPERIMENT:

You will need:	• paper towels	• toothpicks
• blueberry muffin	• paper clips	• pins

Steps:
1. Place the muffin on a paper towel.
2. Describe the surface of the muffin. _____
_____
3. Draw a diagram of the muffin and speculate where the "deposits" (blueberries) are located, both on the surface and inside the muffin.

4. Using paper clips, toothpicks, and pins, remove the "deposits" from the muffin. Remember that you will have to reclaim your "land." This means the "land" will need to be returned to its original condition after mining. You may have to decide whether to mine some deposits.
5. When you have mined as much as possible, lay aside the deposits. Try to return the "land" to its original state.
6. Describe the surface of the mined muffin. _____
_____

ANALYZE AND CONCLUDE:
1. Compare the condition of the muffin before and after mining. How did the surface areas change as a result of mining? _____
_____

© *Earth and Space Science* • Natural Resources

Lab 6.1.4B Mining Conditions

2. What problems were encountered when trying to reclaim the land? _____
_____
_____

3. Relate these problems to the problems that mining companies face. _____
_____
_____

4. What factors should determine if mining should be allowed in areas where the land cannot be reclaimed? _____
_____
_____
_____

5. Were portions of the land mined only to reveal that no deposit lay underneath? _____

6. Were some deposits too difficult to reach without destroying the land? _____
_____

7. Is it practical to refrain from mining anything from the earth? _____
_____
_____
_____
_____

8. What solutions can you suggest to lessen the land destruction caused by mining?
_____
_____
_____
_____
_____
_____

© *Earth and Space Science • Natural Resources*

Name: _____ Date: _____

Renewable and Nonrenewable Resources   WS 6.1.1A

Color of Renewable Resource Beans: _____

Color of Nonrenewable Resource Beans: _____

The beans that represent the renewable resources may be returned to the jar after they have been drawn. Renewable beans should be included in the count of beans remaining in the container each year. Ten beans will be drawn the first year for each scenario.

Scenario 1: No population growth. Constant amount of energy used annually.

1. Predict how many years it will take for the nonrenewable beans to be completely depleted. _____

2. Individual wearing the blindfold, select 10 beans from the jar to represent energy use for one year.

3. After renewable beans have been returned to the jar, how many beans remain in the jar after one year? _____

4. Determine the percentage of nonrenewable beans left in the jar. _____

5. Repeat Steps 2 through 4 until the jar does not contain any more nonrenewable beans. Fill in the table below. If additional years are needed, include the data on a separate sheet of paper.

Scenario 1: No Population Growth									
	Year 1	Year 2	Year 3	Year 4	Year 5	Year 6	Year 7	Year 8	Year 9
# of Beans Drawn	10								
# of Beans Remaining									
% of Nonrenewable Beans									

Scenario 2: Moderate population growth.

1. Predict how many years it will take for the nonrenewable beans to be completely depleted in a growing society. _____

2. Individual wearing the blindfold, select 10 beans from the jar to represent energy use for one year.

3. After renewable beans have been returned to the jar, how many beans remain in the jar after one year? _____

© Earth and Space Science • Natural Resources

Renewable and Nonrenewable Resources

WS 6.1.1A

4. Calculate the percentage of nonrenewable beans left in the jar. _____

5. Determine the increase in rate of consumption for years 2–9 (for example, five additional beans drawn each year).

6. Repeat Steps 2 through 4 until the jar does not contain any more nonrenewable beans. Fill in the table below. If additional years are needed, include the data on a separate sheet of paper.

Scenario 2: Moderate Population Growth									
	Year 1	Year 2	Year 3	Year 4	Year 5	Year 6	Year 7	Year 8	Year 9
# of Beans Drawn	10								
# of Beans Remaining									
% of Nonrenewable Beans									

7. Was your predicted value in Step 1 for both Scenarios 1 and 2 in agreement with the actual results? _____

8. Think of ways to extend energy resources for a growing society. _____

_____
_____
_____
_____
_____
_____
_____
_____
_____
_____
_____
_____

© Earth and Space Science • Natural Resources

Name: _____ Date: _____

Computer Energy Audit   WS 6.1.1B

You will determine how much energy is used in a computer lab or classroom in a 24-hour period. The audit will examine computer monitors, computer CPUs, and printers.

Computer Monitors

Quantity: _____

Determine the amount of energy used by each monitor. This amount can be found on a UL sticker that is located on the back of a computer monitor. The energy usage is measured in watts.

Total energy for one monitor: _____ W

Total energy for all monitors: _____ W

If there are different types of monitors, perform the above calculations for each type in the space below and determine the total amount of energy used.

An average computer monitor uses about 30% of its rated energy usage. Calculate the actual amount of energy used by all of the monitors.

Actual amount of energy used: _____ W

Consider how many hours the monitors are used. Calculate the total amount of energy used each day.

Total energy used each day: _____ W
Convert watts to kilowatts by dividing the above value by 1,000.

Total energy used each day: _____ kW

Computer CPUs

Quantity: _____

Determine the amount of energy produced by each CPU. This amount can be found on a UL sticker that is located on the back of a computer CPU. Note: There may not always be a separate monitor and CPU. These types are combined units. There will only be one UL sticker.

Total energy for one CPU: _____ W

Total energy for all CPUs: _____ W

If there are different types of CPUs, perform the above calculations for each type in the space below and determine the total amount of energy used.

© *Earth and Space Science* • Natural Resources

Computer Energy Audit

WS 6.1.1B

An average computer CPU uses about 30% of its rated energy usage. Calculate the actual amount of energy used by all of the CPUs.

Actual amount of energy used: _____ W
Consider how many hours the CPUs are used. Calculate the total amount of energy used each day.

Total energy used each day: _____ W
Convert watts to kilowatts by dividing the above value by 1,000.

Total energy used each day: _____ kW

Printers

Quantity: _____
Determine the amount of energy produced by each printer.

Total energy for one printer: _____ W

Total energy for all printers: _____ W
If there are different types of printers, perform the above calculations for each type in the space below and determine the total amount of energy used.

An average computer printer uses about 30% of its rated energy usage. Calculate the actual amount of energy used by all of the printers.

Actual amount of energy used: _____ W
Consider how many hours the printers are used. Calculate the total amount of energy used each day.

Total energy used each day: _____ W
Convert watts to kilowatts by dividing above value by 1,000.

Total energy used each day: _____ kW

Total Energy used in the classroom or computer lab
Add the total number of kilowatts used each day by computer monitors, computer CPUs, and printers.

Total Energy: _____ kW

1. What energy saving practices have already been implemented at the school?
_____
_____

2. List ways the total amount of energy can be reduced in one day. _____
_____
_____

© Earth and Space Science • Natural Resources

Name: _____ Date: _____

Paper, Plastic, or Cloth

WS 6.1.2A

Examine three grocery bags: paper, plastic, and cloth.

	Paper	Plastic	Cloth
What resource is used to make it? Is the resource renewable or nonrenewable?			
How does the growing, mining, manufacturing, or harvesting benefit or harm the environment?			
Are nonrenewable resources used at any point in its manufacturing? How?			
How many times can it be reused?			
Can it be composted (will it decompose in the soil as organic matter) instead of being placed in a landfill?			

© Earth and Space Science • Natural Resources

Name: _____ Date: _____

Lab 6.2.1A Sanitary Landfill

QUESTION: Will the garbage pollute the water?

HYPOTHESIS: _____
_____

EXPERIMENT:

You will need:		
• 1 gal milk jug or 2 L bottle with cap • clay • litmus paper • soil	• graduated cylinder • small pieces of garbage (food scraps, paper clips, aluminum foil, rubber bands, paper, pennies, newspaper, plastic scraps)	• filter • cup or box large enough to support jug • 4 beakers • hot plate

Steps:
1. Turn the jug upside down.
2. Spread a thin layer of clay into the capped portion of the jug.
3. Mix food scraps, paper, and other garbage together. Place the mixture on top of the clay.
4. Sprinkle 2 cm of soil over the garbage mixture.
5. Prop the jug cap-side down into a cup or box.
6. Use the graduated cylinder to pour 25 mL of water over the mixture in the jug. Record your observations.
   _____
   _____

7. Record your observations again after 24 hours. _____
   _____

8. After another 24 hours, test the pH of a sample of clean water. This is the control. Record the pH. _____

9. Pour another 25 mL of water over the mixture. Uncap the jug, and allow the water to drain through the clay into a beaker. Test the pH of the water. Record the pH level. _____

10. Pour half of the water through a filter, and let it drain into a second beaker. Record your observations. _____

11. Pour the remaining water into another beaker, and boil it until the water is gone. As a control, boil the same amount of clean water in another beaker until it is gone. Record your observations. Are there any dissolved solids? _____
    _____

© Earth and Space Science • Pollution Solutions

Lab 6.2.1A Sanitary Landfill

ANALYZE AND CONCLUDE:

1. Which garbage items have decomposed? Why? _____

2. How does the garbage smell? What does this suggest? _____

3. What is the pH of the clean water? _____
   What is the pH of the garbage water? _____

4. How can a leaking sanitary landfill present a problem for the surrounding community? _____

5. Why would putting hazardous materials such as chemicals and oil into a landfill cause a problem? _____

6. What is the purpose of a landfill liner? Is it foolproof? _____

7. What are some alternatives to landfills? _____

8. Consider the advantages and disadvantages of each alternative. Which is the best alternative? Why? _____

Name: _____ Date: _____

Lab 6.2.2A Air Pollution and Plants

QUESTION: How will car exhaust affect bean growth?

HYPOTHESIS: _____
_____

EXPERIMENT:

You will need:	• paper towels	• stopwatch
• lima beans	• marking pencil	
• 3 wide-mouthed jars with lids	• automobile and adult	

Steps:
1. Fold each paper towel 3–4 times to fit in the bottom of a jar. Saturate each one with water and place one in the bottom of each jar.
2. Place about 5 beans in each jar.
3. Cap one of the jars. Label it *Control*.
4. Have an adult start a car and expose the second jar to the exhaust fumes for 30 seconds. (Be careful not to breathe in any of the fumes.) Cap the jar immediately to trap the gases. Label this jar *Air Pollution—30 seconds*.
5. Expose the third jar to the exhaust fumes for two minutes. Cap the jar immediately to trap the gases. Label this jar *Air Pollution—2 minutes*.
6. Illustrate and describe your observations of the beans after 2 days, 3 days, and 5 days. Estimate the height of the tallest sprout in each jar and record this data.

Time	Control	Air Pollution—30 seconds	Air Pollution—2 minutes
2 days			
Height			
3 days			
Height			
5 days			
Height			

© *Earth and Space Science* • Pollution Solutions

Lab 6.2.2A Air Pollution and Plants

ANALYZE AND CONCLUDE:

1. Which set of beans grew the tallest? _____

2. Which set of beans looked the healthiest? Were they also the tallest? _____
_____

3. To what type of pollution did you expose the second and third jars? _____
_____
_____

4. Does the length of time that the seeds were exposed to pollutants affect growth? Why? _____
_____
_____
_____

5. Do you think this lab accurately tests the harmful effects of air pollution? Why?
_____
_____
_____
_____
_____

6. Create a new experiment to test the harmful effects of air pollution on plants. Be sure to have a control for comparison and only one variable to test.
_____
_____
_____
_____
_____
_____
_____
_____

Name: _____ Date: _____

Lab 6.2.2B The Air We Breathe

QUESTION: What types of airborne pollutants are in the school?

HYPOTHESIS: _____
_____

EXPERIMENT:

You will need:	• glue	• string
• scissors	• clear packing tape	• microscope
• heavy paper	• hole punch	

Steps:
1. Cut the three holes in the collector card as marked. Glue the card to a heavy piece of paper. Cut out the holes.
2. Tape a piece of packing tape across the back of the card so the adhesive shows through all the holes. Be careful not to touch the adhesive showing through the holes.
3. Place the card on the microscope to observe the adhesive. Illustrate what you see.

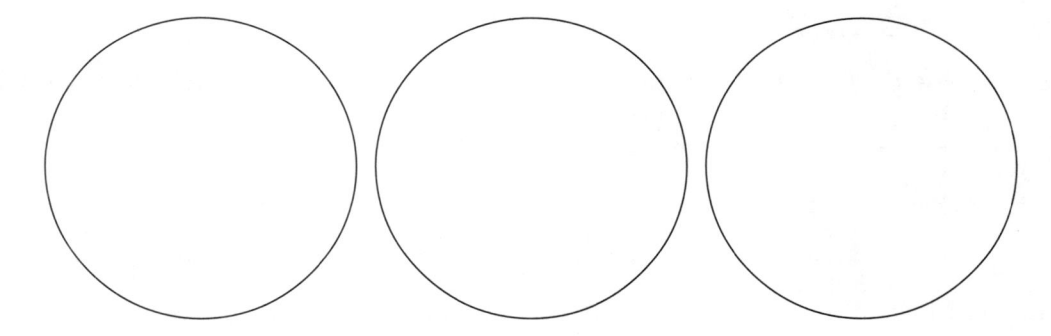

◯ ◯ ◯

4. Write your name, location where you will hang the card, the date, and the time.
5. Punch a hole in the top. Use the string to hang the card somewhere in the school.
6. After a week, retrieve the card and look at the adhesive under the microscope.

What do you see? Illustrate your results. _____

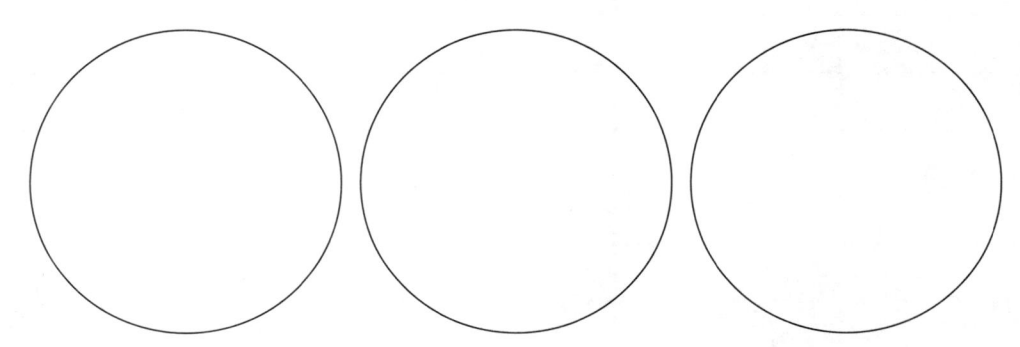

◯ ◯ ◯

© Earth and Space Science • Pollution Solutions

Lab 6.2.2B The Air We Breathe

ANALYZE AND CONCLUDE:

1. Does the tape look different than when you started? _____

2. What types of particulate matter do you think you see on the tape? _____
_____
_____
_____

3. Using the information in the textbook and other research, hypothesize what types of air pollution was collected. Explain. _____
_____
_____
_____
_____

4. What could be done to reduce the air pollution in the room or area you tested?
_____
_____
_____
_____
_____
_____
_____
_____
_____
_____
_____

Name: _____   Date: _____

Lab 6.2.3A Oil Spill

QUESTION: How does an oil spill affect waterfowl?

HYPOTHESIS: _____

_____

EXPERIMENT:

You will need:	• paper towels	• bowl
• shallow tub	• vegetable oil	• 2 plastic straws
• feather	• dish soap	• cotton balls
• blue food coloring	• sponge	• syringe or pipette

Steps:
1. Fill the tub halfway with water. Add 3 drops of food coloring.
2. Set the feather in the water. What happens? _____
3. Remove the feather and set it on paper towels to dry.
4. Pour 40 mL of oil in the water. What does the oil do? _____
_____
5. Place the feather on top of the oil, holding one end. Gently move the feather around to coat it with oil. Does it float? Describe how the feather looks and feels.
_____
_____
_____
6. Remove the feather and set it on the paper towels. Put water in the bowl and wet the sponge. Add a drop of dish soap to the sponge and clean the feather.
7. When finished cleaning, rinse the feather and let it dry.
8. Take the straw(s) and cut or connect them so their measurement equals the width of the tub.
9. Place the straw at one end of the pan and carefully push the oil toward the other end.
10. On the side of the straw that has only a little oil left, use a cotton ball to absorb the oil.
11. Use the syringe to remove the large patch of oil.

ANALYZE AND CONCLUDE:

1. Does the feather feel as light as it did at the beginning of the experiment? _____

2. How well do you think the feather would float after being soaked in oil and cleaned? _____

© Earth and Space Science • Pollution Solutions

Lab 6.2.3A Oil Spill

3. How do you think water fowl are affected by oil spills? _____
_____
_____

4. Were you able to remove all the oil from the water? Why? _____
_____
_____
_____

5. What are some other ways you could remove the oil from the water? _____
_____
_____
_____
_____

6. Do you think oil spills in the ocean are more difficult to clean than the oil spill in the experiment? Why? _____
_____
_____
_____

7. Oil burns and water does not. Would burning the oil be a good solution to get rid of it? Why? _____
_____
_____
_____
_____
_____
_____
_____

© Earth and Space Science • Pollution Solutions

Name: _____   Date: _____

Lab 6.2.3B Acid Rain

QUESTION: How does acid rain affect plants?

HYPOTHESIS: _____

_____

EXPERIMENT:

You will need:	• masking tape	• white vinegar
• 3 young bean plants	• marking pen	• water
• 3 jars with lids		

Steps:
1. Label each plant: *A*, *B*, or *C*.
2. Label each jar: *A*, *B*, or *C*.
3. Add 120 mL of white vinegar to Jar A.
4. Add 60 mL of white vinegar to Jar B.
5. Do not add any vinegar to Jar C.
6. Add 120 mL of water to each jar and stir gently to mix the solution.
7. Set the bean plants in a sunny location.
8. Water each plant with 60 mL of solution from its corresponding jar.
9. Record observations daily for 5 days. If the soil gets dry, water again with 60 mL of solution. Record any additional watering.

Day	Plant A	Plant B	Plant C
Start, 1			
2			
3			
4			
5			

© Earth and Space Science • Pollution Solutions

Lab 6.2.3B Acid Rain

ANALYZE AND CONCLUDE:

1. Which plant is the healthiest? Why? _____

2. Which plant is the least healthy? Why? _____

3. Were your results what you expected? Why? _____

4. How is the experiment similar to acid rain? _____

5. What causes acid rain? _____

Name: _____ Date: _____

Recycling Journal WS 6.2.1A

Record the items you throw away and recycle. Do this for two days.

Trash

Throw Away	Recycle

1. Which did you do more of—throwing away or recycling? _____

2. Were you surprised at how much trash you generate? Why? _____
_____
_____

3. In what ways could you recycle more and throw away less? _____
_____
_____
_____
_____

© Earth and Space Science • Pollution Solutions

Name: _____ Date: _____

Do the Math
WS 6.2.4A

Sometimes an individual effort seems insufficient. But do the math for the following examples, and you will see how thinking globally and acting locally can make a difference.

1. An energy-conserving light bulb costs about three times as much as a regular light bulb, but the conserving light bulbs last four times longer and are a fourth of the cost of a regular bulb to operate. If an energy-conserving light bulb costs $6.00, lasts 24 months, and costs $0.35/day to operate, what would be the cost to purchase and operate a regular light bulb for the same time frame? Which light bulb costs more to use over 24 months? How much more?

2. Running the water while brushing your teeth could waste about 550 mL of water. If you and your family members turned off the water while brushing your teeth for one whole year, how much water could you save? How many times each day does each member brush? How much water could your family save each year? Add your answer to all your classmates' answers. How much water could the whole class save each year?

3. How many paper cups do you use in a week? Each time you use a paper cup and throw it away, you are adding to the landfill. Add the number you use to all that your family uses. How many paper cups does your family use each year? How many cups are used by all the families in your class each year? Write down some ways to reduce the number of paper cups you and your family use.

© Earth and Space Science • Pollution Solutions

Do the Math

WS 6.2.4A

4. Determine how many trees your school could save each year by recycling paper. Make a plan to collect all paper trash from every classroom, office, and teacher workroom for a week. Collect the paper trash and determine its mass in kilograms. For every 910 kg of paper recycled, 17 trees could be saved. Take the mass of the paper you collected, multiply that by the number of weeks in a school year. This is the number of kilograms you can recycle. Use the ratio of 910 kg/17 trees to determine how many trees your school could save each year if it recycled paper.

5. Is reducing, reusing, and recycling a good thing to do? Why? _____
_____
_____
_____
_____

6. How can you reduce, reuse, and recycle? _____
_____
_____
_____
_____
_____
_____

Name: _____ Date: _____

Lab 7.1.2A Sundial

QUESTION: How can shadows be used to tell time?

HYPOTHESIS: _____
_____

EXPERIMENT:

You will need:	• metric ruler	• flashlight
• paper plate	• protractor	

Steps:
1. Draw a line through the approximate center of the paper plate using the ruler.
2. Use the protractor and ruler to mark and draw a line that runs through the center of the first line and is perpendicular to it. The lines should create a giant plus sign that divides the plate into four equal sections. Label each line *North*, *South*, *East*, or *West*, so the plate looks like a simple magnetic compass without a needle.
3. Mark 30° segments around the plate using the protractor. If you live in the Northern Hemisphere, label the *North* line *12*. If you live in the Southern Hemisphere, label the *South* line *12*. If you live near the equator, you may write *12* on either the *North* or the *South* line. Continue numbering each line so the plate looks like a clockface.
4. Carefully poke a pencil through the center of the plus sign and adjust it so it stands straight up.
5. In a darkened room, use a flashlight to make the pencil's shadow appear on the plate.
   - Make the pencil's shadow as long and as short as possible. Observe the flashlight's positions when the shadow is longest and shortest.
   - Make the pencil's shadow move clockwise around the marks on the plate. Observe the flashlight's position in relation to the shadow.
   - Shine the flashlight on the pencil from different angles. Observe how the shadow changes as the angle of the light changes.

ANALYZE AND CONCLUDE:

1. How was the flashlight positioned to create the shortest pencil shadow? _____
_____
_____

2. How was the flashlight positioned to create the longest pencil shadow? _____
_____
_____

Lab 7.1.2A Sundial

3. Why should a sundial be situated so the mark for noon points north in the Northern Hemisphere and points south in the Southern Hemisphere? _____

_____
_____
_____
_____
_____

4. At the equator, what would a sundial's shadow look like? _____

_____
_____

5. How might the length of shadows change at different times during the year?

_____
_____
_____
_____
_____

6. How could shadows be used to help calculate the length of a solar year? _____

_____
_____
_____
_____
_____

7. What challenges exist when using shadows to help measure a solar year? _____

_____
_____
_____
_____
_____

Name: _____ Date: _____

Lab 7.1.3A Atmospheric Gases

QUESTION: How do the gases in the atmosphere affect people's view of the universe?

HYPOTHESIS: _____
_____

EXPERIMENT:

You will need:	• tape	• hot plate
• straight pin	• 2 books	
• aluminum foil	• flashlight	

Steps:
1. Use the straight pin to poke about 20 holes in the aluminum foil.
2. Stand the books upright on a flat surface and tape the aluminum foil between them so light comes through the holes.
3. Darken the room and shine a flashlight through the holes. Record your observations. _____
_____
_____

4. Plug in a hot plate and allow it to heat up. Place the aluminum foil in front of the hot plate.
5. Shine the flashlight over the hot plate through the holes. Record your observations. _____
_____
_____
_____

ANALYZE AND CONCLUDE:
1. What differences did you notice about the light coming through the holes with and without the heat source? _____
_____
_____
_____
_____

© Earth and Space Science • Solar System

Lab 7.1.3A Atmospheric Gases

2. How can you explain the differences between the light coming through the holes with and without the heat source? _____

3. Why do you think stars appear to twinkle? _____

4. Explain whether you think stars would appear to twinkle if you saw them from the moon. _____

5. Explain whether you think the twinkling of stars helps or inhibits scientific observation. _____

Name: _____ Date: _____

Lab 7.1.3B Refracting Telescope

QUESTION: How does a refracting telescope work?

HYPOTHESIS: _____

_____

EXPERIMENT:

You will need:	• convex lens	• aluminum foil
• concave lens	• scissors	• straight pin

Steps:
1. Select a well-lit, distant object to view through the telescope.
2. Hold the concave lens close to one eye. This lens is your eyepiece.
3. Hold the convex lens directly in front of the eyepiece. (The concave lens should be between your eye and the convex lens.) The convex lens is the objective lens.
4. Slowly move the objective lens in a straight line away from the eyepiece until the object you are viewing comes into view through both lenses. Once you are able to see the object, move the objective lens in and out until you can see your object clearly. You may need a partner to move the objective lens out farther than your arm can reach.
5. Compare how the image appears when viewed with and without lenses. Note changes in the image's clarity, brightness, and size, and if it was inverted.

_____
_____
_____
_____

6. Cut a square of aluminum foil measuring about 10 cm × 10 cm.
7. Use the straight pin to make a tiny hole approximately in the center of the aluminum foil square.
8. Repeat Steps 1–4 but use the foil square's pinhole as your eyepiece.
9. Compare how the image appears when viewed with and without lenses. Note changes in the image's clarity, brightness, and size, and if it was inverted.

_____
_____
_____
_____

Earth and Space Science • Solar System

Lab 7.1.3B Refracting Telescope

ANALYZE AND CONCLUDE:

1. Compare your observations using the lens and foil pinhole as eyepieces.
   _____
   _____

   Clarity: _____
   _____
   _____
   _____

   Brightness: _____
   _____
   _____
   _____

   Size: _____
   _____
   _____
   _____

   Inversion: _____
   _____
   _____
   _____

2. Using what you observed, what is the eyepiece's purpose? _____
   _____
   _____

3. Using what you observed, what is the purpose of the objective lens? ____
   _____
   _____
   _____

4. What do you think accounts for the differences between the way the lens and the foil pinhole function as eyepieces? _____
   _____
   _____
   _____

Name: _____   Date: _____

Lab 7.1.5A Indirect Evidence

QUESTION: How helpful is indirect evidence in drawing conclusions about an object?

HYPOTHESIS: _____
_____

EXPERIMENT:

You will need:	• triple beam balance	• bar magnet
• 4 clay balls from your teacher	• masses for balance	
• 4 pushpins of different colors	• metric ruler	

Steps:
1. Place a different colored pushpin into each planet to identify it.
2. Use a triple beam balance and masses to determine the mass of each planet in grams. Record the data.
3. Use a metric ruler and the formula to the right to determine the volume of each planet in cubic centimeters. Record the data. $V = \dfrac{4}{3\pi r^3}$
4. Use the formula Density = mass/Volume to the right to calculate each planet's density in g/cm³. Record the data. $D = \dfrac{m}{V}$
5. Use the magnet to test each planet for magnetic properties.
6. Using the information you gathered, hypothesize what is at the core of each planet. Explain your reason for each hypothesis.

	A	B	C	D
Pushpin Color				
Mass				
Volume $V = \dfrac{4}{3\pi r^3}$				
Density $D = \dfrac{m}{V}$				
Magnetic?				
Hypothesis and Explanation				

7. Deconstruct the clay balls to reveal their cores.

© Earth and Space Science • Solar System

Lab 7.1.5A Indirect Evidence

ANALYZE AND CONCLUDE:

1. What difficulties did you encounter when trying to determine the core of each planet? _____
_____
_____
_____
_____
_____
_____
_____

2. How accurate were your hypotheses about each planet? _____
_____
_____

3. How accurate was your initial hypothesis in response to the question, "How helpful is indirect evidence in drawing conclusions about an object"? _____
_____
_____

4. Revise your initial hypothesis if necessary and support it by referring to your experiences in completing this lab. _____
_____
_____
_____
_____
_____
_____
_____

© Earth and Space Science • Solar System

Name: _____ Date: _____

Lab 7.1.6A Diameter of the Sun

QUESTION: What is the sun's diameter?

HYPOTHESIS: _____
_____

EXPERIMENT:

You will need:	• pushpin	• tape
• 2 white unruled index cards	• meterstick	

WARNING: Never look directly at the sun.

Steps:
1. Fold a third of one index card lengthwise.
2. Use the pushpin to make a pinhole in the center of the large section of the folded index card.
3. Tape the smaller section of the card to the end of the meterstick so the fold is lined up with the stick's 0 mark. The folded side with the pinhole in its center should be perpendicular to the stick.
4. On the center of the second index card, draw two parallel lines 0.7 cm apart. This measure will be the width of the sun's image.
5. Position the meterstick so the sun's rays pass through the pinhole.
6. Move the second card on the meterstick so an image of the sun appears.
7. Move the second card along the meterstick until the sun's image just fits between the 0.7 cm marks.
8. Record the distance between the two cards. _____
_____
9. Write the proportion of the size of the image to the distance between the cards.
_____
10. Write the proportion of the sun's diameter to the distance between Earth and the sun. Let d represent the sun's diameter. The approximate distance between Earth and the sun is 150,000,000 km. _____
_____
11. Solve the proportion for d.

© Earth and Space Science • Solar System

Lab 7.1.6A Diameter of Sun

ANALYZE AND CONCLUDE:

1. Why do you think your calculations do not exactly match the actual diameter of the sun? _____

2. Do you think it is possible to obtain a perfectly accurate measurement of the sun's diameter? Why? _____

3. How accurate do you think measurements are of the sun's layers? Why? _____

© Earth and Space Science • Solar System

Name: _____ Date: _____

Galileo and the Sun   WS 7.1.1A

Which of the following statements accurately describes what was happening at the time of Galileo with regard to science and religion? On the line following each statement, indicate whether the statement is *true*, *false*, or you *do not know*.

1. Galileo was a scientist who was persecuted by the church for his scientific findings. _____

2. Galileo was a scientist who got into trouble for attacking the Pope. _____

3. At the time of Galileo, the church believed in geocentrism and scientists believed in heliocentrism. _____

4. At the time of Galileo, most scientists as well as theologians believed in geocentrism. _____

5. Heliocentrism was a controversial new theory that the church rejected because it went against the Bible. _____

6. Heliocentrism was uncontroversial and seen as a possibility, but it had not yet been proven and some scientific arguments counted against it. _____

7. The Galileo story is all about having to choose between religion and science. _____

© Earth and Space Science • Solar System

Name: _____   Date: _____

Galileo Goes to Jail　　　　　　　　　　　　　　　　　　　　　　　WS 7.1.1B

What do you know about Galileo? And what have you heard about his role in the history of the relationship between science and religion? The imprisonment and torture of the scientist Galileo (1564–1642) by the Church for advocating Copernicanism and replacing old religious dogma with true scientific knowledge is a widely told story. But is the popular version really true to the historical events?

What really happened with Galileo? And why might such a rumor persist despite historical evidence to the contrary? When Galileo perfected his newly developed telescope, it gave him access to data unavailable to other scientists. Simply put, he could see things that others could not see. As a result, Galileo became convinced of Copernicus' theory of the earth's movement around the sun. In this he was not going along with widely accepted science in the face of inflexible theology. The data supporting his views was inaccessible to others, and went against widely accepted scientific views of the day, even though it eventually proved to be correct.

Other scientists, lacking physical evidence and suspicious of data coming only from the one source provided by this newfangled technology, were hesitant to overturn both accepted scientific theory and a traditional understanding of Scripture on such an uncertain basis. In fact, they attacked Galileo's ideas because they were unable to corroborate them. When church leaders cast doubt on his views, they were echoing the views of most scientists at the time. Moreover, they did not reject the theory out of hand, but declared it a possibility rather than an established fact. It is perhaps hardly surprising that theologians did not immediately rush to definitively overturn the consensus view based on one person's findings with a new device. The idea of heliocentrism itself was relatively uncontroversial by this point; Pope Gregory XIII had used it reform the calendar. The theory was not being opposed by church institutions, but it was still unproven.

In response, Galileo brashly refuted the biblical arguments against Copernicanism, wading into theological discussions in a manner that attacked the Pope, writing a dialog in which the Pope's words were put in the mouth of a character called Simplicio (simpleton). This alienated him from former Catholic church allies and lost him the Pope's earlier support. In a post-Reformation era where the Church's authority was regularly under attack, lay peoples' reinterpretation of Scripture was subjected to extra scrutiny. Galileo's way of interpreting the Bible went outside authorized channels and smacked of Protestantism. He got into trouble more because of politics than science. Although his publications contained neither a categorical assertion of Copernicanism nor a denial of the authority of Scripture, Galileo's way of approaching the Bible and outspoken approach led to his being charged with "suspicion of heresy." Eventually, under ecclesial pressure, he retracted his advocacy of Copernicanism.

For his views, the Church sentenced Galileo to "formal imprisonment" and "rigorous examination," and widely publicized this as a deterrent to others. This led many to envision Galileo suffering torture and rotting in an inquisition prison. But because of Galileo's popularity and a measure of ecclesial good will, he instead served nine

Galileo Goes to Jail

WS 7.1.1B

years in comfort under house arrest. During this time, he continued to prepare scientific writings. His imprisonment and torture never happened. They survive only as myth—even after relevant documents came to light showing that Galileo had suffered neither.

Why was it such a popular move to exaggerate what Galileo suffered? The story seemed to pit a backward and superstitious Church against new, brave scientific luminaries, and this was a useful storyline for those supporting the idea of conflict between faith and science.

It takes attentiveness to historical detail to acknowledge the complexity of the Galileo affair, and cultural debates are often conducted in more simplistic terms. Galileo's punishment stemmed from a variety of interconnected factors: a series of disputes about novel and inaccessible scientific data, the development of a new technology, the personal offense caused by Galileo's brash response, and his bold claims in the field of biblical interpretation, in which he lacked authority.

1. Review the responses you made on **WS 7.1.1A Galileo and the Sun**. Are there any answers you would change after reading the above article? Why? _____
_____
_____

2. What does Galileo's story teach about how people should approach the relationship between science and faith? _____
_____
_____
_____

3. How might it have required courage, humility, and patience for Christians who were not scientists, scientists who were not Christians, and Christians who were scientists to think through these issues at the time of Galileo? _____
_____
_____
_____

Name: _____ Date: _____

Astronomical Observations   WS 7.1.2A

Ancient people did not have watches, clocks, or calendars, but they did have reliable methods of keeping track of time. These methods used careful observations of the sky and shadows. Answer the following questions to discover how aware you are of astronomy:

Short Answers

1. During which season are shadows the longest? _____

2. How do shadows change throughout the day? _____

3. When was the last full moon? _____

True or False

Indicate whether the following statements are true or false. Explain why the false statements are false.

1. When the moon is crescent-shaped, it points in the same direction. _____

2. The moon is always visible on a cloudless night. _____

3. Some days of the year predictably have more meteorites than others. _____

4. Stars are all the same color. _____

5. Other planets are visible from Earth. _____

6. The stars look the same in the sky every night. _____

7. People can see other galaxies without a telescope. _____

8. Certain artificial satellites that are orbiting Earth are visible without a telescope.

© Earth and Space Science • Solar System

Name: _____   Date: _____

Solar System Diagram

WS 7.1.5A

Label the *sun*, the *terrestrial planets*, the *asteroid belt*, the *gas giants*, the *dwarf planet*, and the *Oort cloud*. This diagram is not drawn to scale.

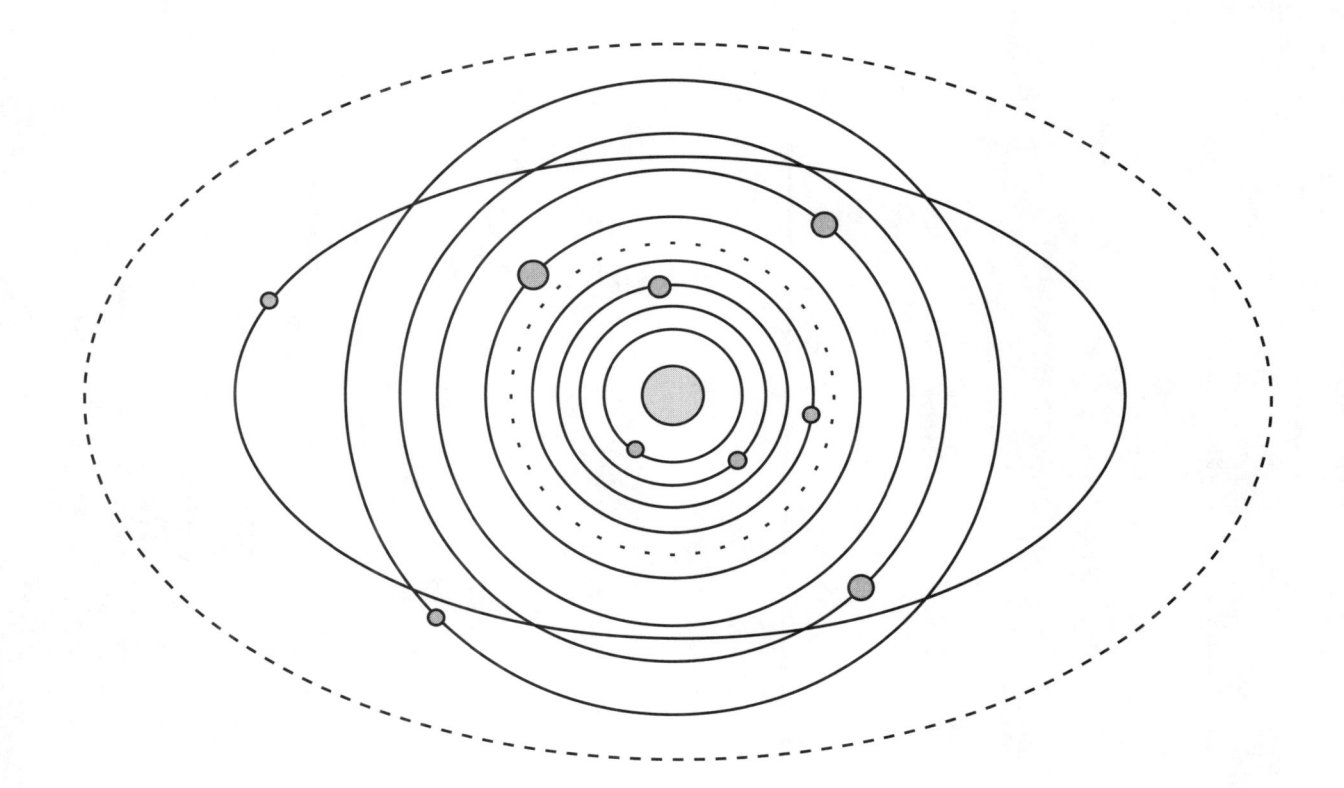

© Earth and Space Science • Solar System

Name: _____ Date: _____

Sun Diagram

WS 7.1.6A

Label each layer of the sun and list a specific fact about each layer.

© *Earth and Space Science* • Solar System

Name: _____ Date: _____

Lab 7.2.1A Ellipse

QUESTION: What is an ellipse?

HYPOTHESIS: _____
_____

EXPERIMENT:

You will need:	• cardboard	• 30-centimeter piece of string
• 2 pushpins	• metric ruler	

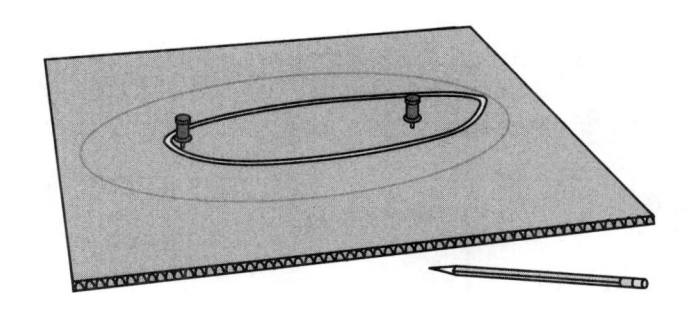

Steps:
1. Stick two pushpins into the cardboard about 10 cm apart.
2. Loop the string around the pushpins and tie the ends together. One person should hold the pushpins in place. Another person should place a pencil inside the string and trace an ellipse; keep the string tight at all times. Label this *Ellipse A*.
3. Move the pushpins until they are 5 cm apart and trace another ellipse. Label this *Ellipse B*.
4. Choose three points on Ellipse A. Measure each point's distance to each focus (pushpin) and add the two distances. Record your measurements in the chart.
5. Choose three points on Ellipse B. Measure each point's distance to each focus (pushpin) and add the two distances. Record your measurements in the chart.

	Point 1 Sum	Point 2 Sum	Point 3 Sum
Ellipse A			
Ellipse B			

© Earth and Space Science • Planets

Lab 7.2.1A Ellipse

ANALYZE AND CONCLUDE:

1. How did the distance between the two foci (fixed points) affect the ellipses' shape?
_____
_____
_____
_____

2. What shape would you draw if you removed one focus? _____
_____

3. What does any point on an ellipse have in common with any other point on an ellipse? _____
_____
_____

4. Why might some of your sums not be identical for all three points on an ellipse?
_____
_____
_____
_____

Use the Internet to find the answers to the questions below.

5. Which planet's orbit is the most elliptical? _____

6. Which planet's orbit is the most circular? _____

Name: _____ Date: _____

Lab 7.2.2A Venus

QUESTION: Why is the temperature of Venus so hot?

HYPOTHESIS: _____

_____

EXPERIMENT:

| You will need: | • 2 thermometers | • rubber band |
| • 2 books | • balloon | |

Steps:
1. Set two books side by side in direct sunlight next to a window.
2. Place a thermometer on each book with the thermometer's bulb facing the windows and overhanging the books by several centimeters.
3. Inflate a balloon and use a rubber band to secure its end over the bulb of one thermometer.
4. Record both thermometers' readings every minute for 10 minutes.

Time	Thermometer: No Balloon	Thermometer: Balloon
1 min		
2 min		
3 min		
4 min		
5 min		
6 min		
7 min		
8 min		
9 min		
10 min		

Lab 7.2.2A Venus

ANALYZE AND CONCLUDE:

1. Explain the differences in temperature. _____
_____
_____
_____
_____
_____
_____
_____

2. Compared to Earth, the temperature of Venus is very hot. Why might this be? ____
_____
_____
_____
_____
_____
_____
_____
_____
_____

3. How might the relationship between Venus's atmosphere and temperature help scientists better understand conditions on Earth relating to the atmosphere and changing temperature trends? _____
_____
_____
_____
_____
_____
_____
_____
_____
_____
_____

Name: _____ Date: _____

Kepler's Third Law   WS 7.2.1A

Planets far from the sun take longer to orbit than planets close to the sun. According to Kepler's third law, the cube of the radius of a planet's orbit is equal to the square of its period of revolution. The radius (a) of each planet is given in astronomical units (AU) on the chart below. The variable P represents a planet's period of revolution in Earth years. Using the formula $a^3 = P^2$, the radius of each planet's orbit, and a calculator, solve for the values of a^3, P^2, and P. Round all answers to two decimal places.

Planets' Period of Revolution

Planet	a	a^3	P^2	P
Mercury	0.39 AU			
Venus	0.72 AU			
Earth	1.00 AU			
Mars	1.52 AU			
Jupiter	5.20 AU			
Saturn	9.58 AU			
Uranus	19.20 AU			
Neptune	30.05 AU			

© Earth and Space Science • Planets

Name: _____ Date: _____

Gravity   WS 7.2.1B

Use Internet resources or an encyclopedia to help you fill in the chart. Round your answers to two decimal places. Write the mass of each planet using exponents. For example, the mass of Earth is 5,972,190,000,000,000,000,000,000 kg. Written with exponents and rounded to two decimal places, it is 5.97×10^{24} kg.

Factors of Gravity

Planet	Density (g/cm³)	Mass (kg)	Gravity (m/s²)
Mercury			
Venus			
Earth		5.97×10^{24}	
Mars			
Jupiter			
Saturn			
Uranus			
Neptune			

1. If a planet's gravity is greater than Earth's, but its density is less than Earth's, how does its mass compare to that of Earth's? _____

_____

2. If a planet's density is about the same as Earth's, but its mass is much less than Earth's, will the planet's gravity be greater than or less than Earth's gravity?

_____

3. If a planet's mass is much greater than Earth's, but its gravity is about the same as Earth's, how must its density compare to Earth's density? _____

_____

4. Explain how density, mass, and gravity are related. _____

_____

_____

_____

© *Earth and Space Science* • Planets

Name: _____ Date: _____

Inner Planets' Properties

WS 7.2.2A

	Period of Rotation in Earth Units	Atmosphere	Surface Temperature (°C)	Distance from Sun (km)	Distance from Earth (km)	Surface Features	Notable Sights
Mercury							
Venus							
Earth							
Mars							

© Earth and Space Science • Planets

Inner Planets' Properties

WS 7.2.2A

	Diameter	How long would it take to fly around the equator at 900 kph?	Percent of Earth's Gravity	If a suitcase weights 20 kg on Earth, how much would it weigh on this planet?	Period of Revolution in Earth Units	How old would you be on this planet?
Mercury						
Venus						
Earth						
Mars						

© Earth and Space Science • Planets

Name: _____ **Date:** _____

Outer Planets' Properties

WS 7.2.3A

	Period of Rotation in Earth Units	Atmosphere	Surface Temperature (°C)	Distance from Sun (km)	Distance from Earth (km)	Surface Features	Notable Sights
Jupiter							
Saturn							
Uranus							
Neptune							

© Earth and Space Science • Planets

Outer Planets' Properties

WS 7.2.3A

	Jupiter	Saturn	Uranus	Neptune
Diameter				
How long would it take to fly around the equator at 900 kph?				
Percent of Earth's Gravity				
If a suitcase weights 20 kg on Earth, how much would it weigh on this planet?				
Period of Revolution in Earth Units				
How old would you be on this planet?				

© Earth and Space Science • Planets

Name: _____ Date: _____

Triple Venn Diagram

WS 7.2.4A

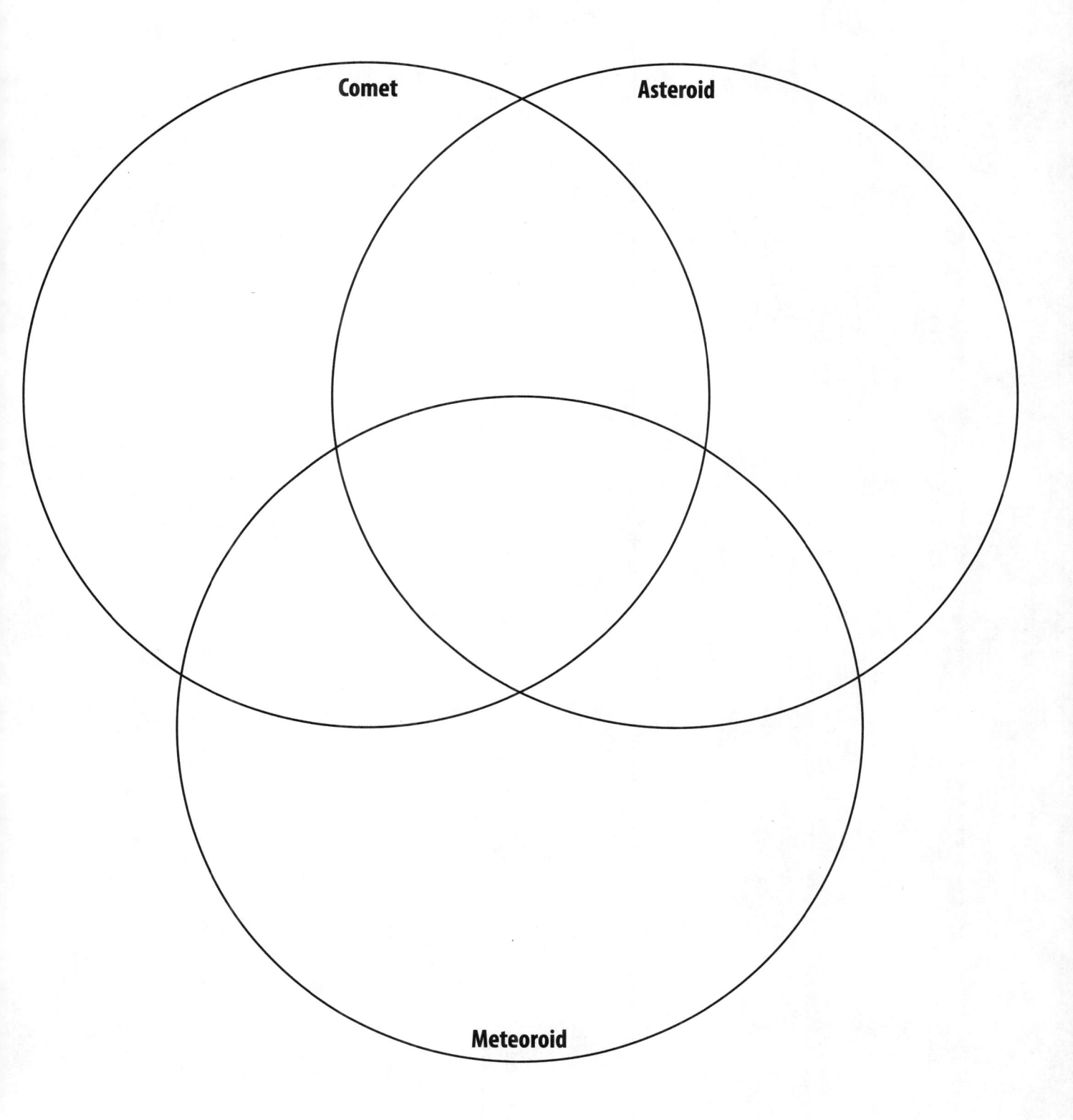

Comet

Asteroid

Meteoroid

© *Earth and Space Science* • Planets

Name: _____ Date: _____

Lab 7.3.1A Moon Width

QUESTION: How wide is the moon?

HYPOTHESIS: _____
_____

EXPERIMENT:

| **You will need:** | • two 10 cm × 7 cm cardboard rectangles | • tape |
| • straight pin | | • meterstick |

Steps:
1. Poke a pinhole into the center of one piece of cardboard.
2. Tape the cardboard to the 0 mark on a meterstick in such a way that the pinhole is visible when you hold the meterstick in front of you.
3. Create a 0.5-centimeter hole in the center of the second piece of cardboard.
4. Hold the meterstick with the 0 mark away from you. Look at the moon through the pinhole in the cardboard taped to the meterstick. Move the second card back and forth along the meterstick until the moon fits exactly inside the 0.5 cm hole.
5. Record the distance in centimeters between the two cards on the meterstick.

_____

ANALYZE AND CONCLUDE:
1. The average distance between Earth and the moon is about 382,500 km. Use the following equation to determine the diameter of the moon to the nearest kilometer: moon's diameter × distance between cards = cut circle's diameter × distance from Earth to moon. _____
_____
_____
_____
_____
_____
_____

© Earth and Space Science • Sun, Earth, and Moon

Lab 7.3.1A Moon Width

2. Compare your answer in Question 1 with answers your classmates obtained. Why might the answers be somewhat different? _____

3. The diameter of the moon is nearly 27.25% the diameter of Earth, which is about 12,756 km. Calculate the actual diameter of the moon to the nearest kilometer.

4. Compare your answers in Questions 1 and 3. How accurate was your measurement? Why might differences exist? _____

5. How could you obtain more accurate results if you repeated the experiment?

© Earth and Space Science • Sun, Earth, and Moon

Name: _____ Date: _____

Lab 7.3.1B Regolith Formation

QUESTION: How does moon regolith form?

HYPOTHESIS: _____
_____

EXPERIMENT:

You will need:	• sandpaper	• fist-sized rock
• 5 slices of toasted wheat bread	• ice cube containing sand	• meterstick
• tray	• 2 slices of toasted white bread	

Steps:

1. Hold a slice of wheat toast above a tray and blow on the toast. Describe what falls into the tray. _____
_____
_____

2. Rub the slice of wheat toast from Step 1 with sandpaper over a tray. Describe what falls into the tray. _____
_____
_____

3. Hold an ice cube that contains sand underneath running water until it melts. Describe the remaining particles. _____
_____
_____

4. Place 2 slices of white toast on top of 4 slices of wheat toast on a tray.

5. Hold a rock 30 cm above the stack of toast and drop it. Describe the bread slices and the crumbs. _____
_____
_____
_____

© *Earth and Space Science* • Sun, Earth, and Moon

Lab 7.3.1B Regolith Formation

6. Drop the rock onto the layers of toast 25 times. Which crumbs are visible at the surface? Why? Compare the thickness of the crumb layers now to the results noted following Step 5. _____

_____
_____
_____
_____
_____
_____
_____
_____
_____

ANALYZE AND CONCLUDE:

1. What processes in this activity represent how regolith is formed on Earth?
_____
_____
_____

2. What processes represent how regolith is formed on the moon? _____
_____
_____
_____

3. What process is not a factor of regolith formation on the moon? Why? _____
_____
_____
_____
_____
_____
_____

Name: _____ Date: _____

Moon Map
WS 7.3.1A

1. When the moon is near the first or last quarter, draw the visible portion of the moon. Include any features you can see with your unaided eyes. Then look at the moon through binoculars or a small telescope and complete the drawing.
2. Consult a labeled map of the moon and identify the more prominent features.
3. Identify the beginning time and ending time of your observations.

Beginning time: _____

Ending time: _____

© *Earth and Space Science* • Sun, Earth, and Moon

Name: _____ Date: _____

Going Through a Phase

WS 7.3.2A

Shade the moons to show the visible portion as seen from Earth. Label each phase.

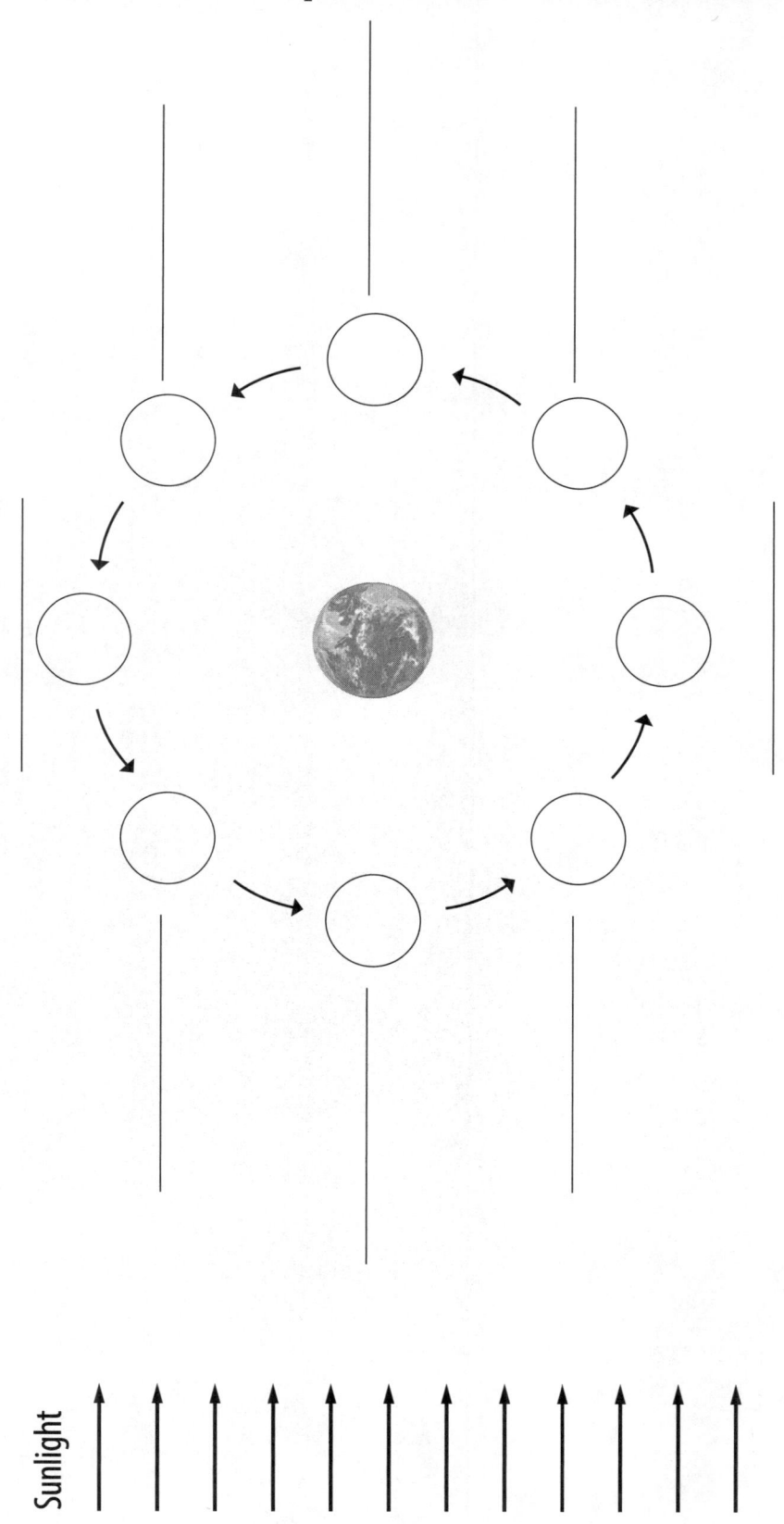

Sunlight

Name: _____ Date: _____

Solar Eclipse Diagram

WS 7.3.3A

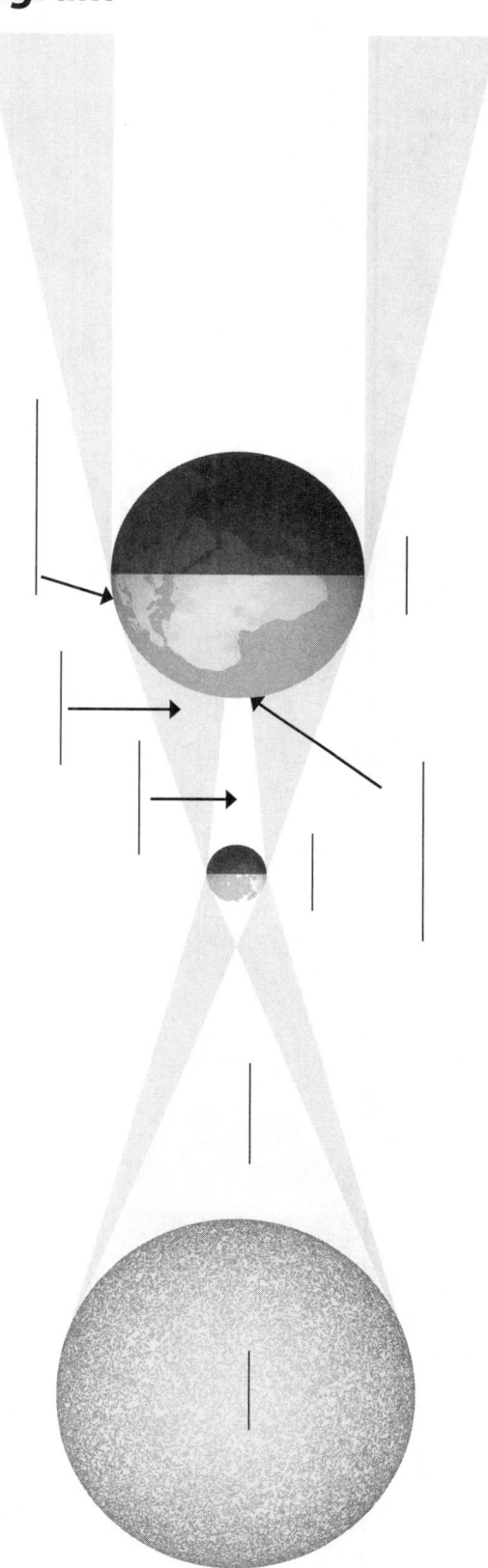

Label the parts of the solar eclipse.

© Earth and Space Science • Sun, Earth, and Moon

Name: _____ Date: _____

Lunar Eclipse Diagram

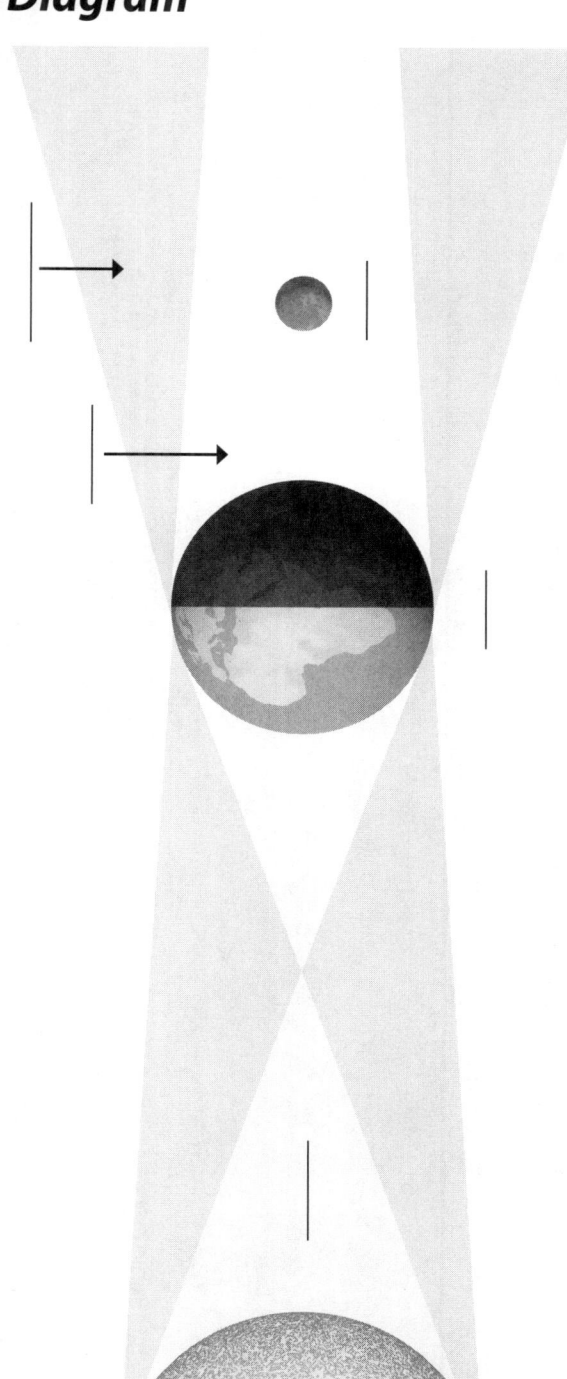

WS 7.3.3B

Label the parts of the lunar eclipse.

© Earth and Space Science • Sun, Earth, and Moon

Name: _____ Date: _____

Changing Seasons

WS 7.3.4A

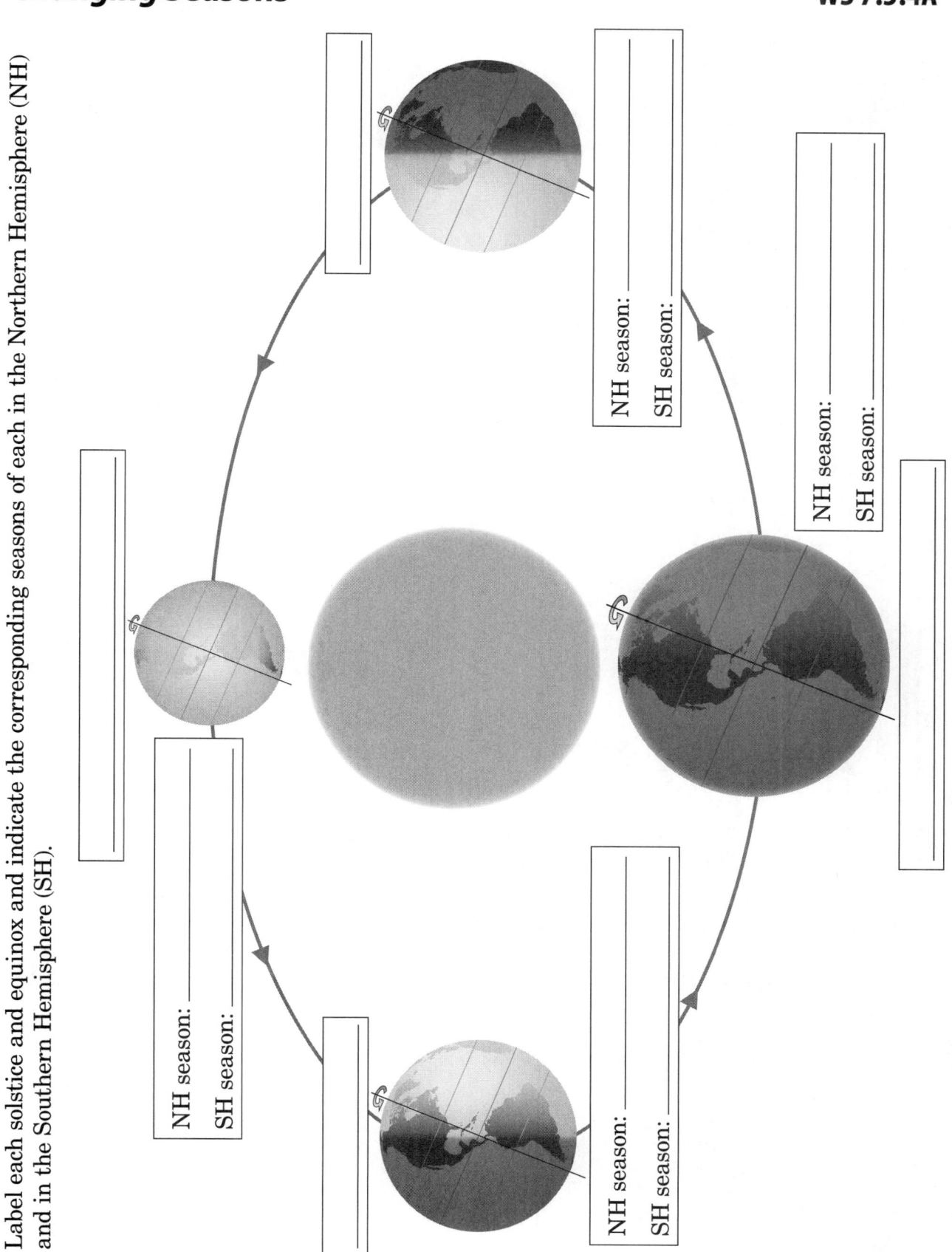

Label each solstice and equinox and indicate the corresponding seasons of each in the Northern Hemisphere (NH) and in the Southern Hemisphere (SH).

NH season: _____
SH season: _____

NH season: _____
SH season: _____

NH season: _____
SH season: _____

NH season: _____
SH season: _____

© Earth and Space Science • Sun, Earth, and Moon

Name: _____ Date: _____

Lab 8.1.1A Star Spectrums

QUESTION: How do scientists determine the elemental components of stars?

HYPOTHESIS: _____

_____

EXPERIMENT:

You will need:	• striker or matches	• baking soda
• large paper clip	• spectroscope	• ground calcium carbonate
• Bunsen burner	• chalk dust	• salt

Steps:
1. Bend a large paper clip to form a small loop on one end.
2. Place the loop over a lit Bunsen burner for a few seconds to burn off contaminants.
3. Moisten the loop and dip it into the chalk dust.
4. With the overhead lights turned off, place the loop and chalk dust over the flame. Using the spectroscope, observe the spectrum produced by the chalk dust. In the box below, illustrate the spectrum produced.

Chalk Dust

5. Repeat Steps 2 through 4 using baking soda, ground calcium carbonate, and salt. In the boxes below, illustrate the spectra observed.

Baking Soda

Calcium Carbonate

Salt

© *Earth and Space Science* • Stars and the Universe

Lab 8.1.1A Star Spectrums

ANALYZE AND CONCLUDE:

1. Compare and contrast the spectra of each substance that was tested. _____

2. Which spectra were the most similar? _____

3. What might the spectra indicate about the substances? _____

4. What information could scientists get from studying the spectra of stars? _____

Name: _____ Date: _____

Parallax WS 8.1.1A

Stars that are closer to Earth appear to shift more than stars that are farther away. To determine the distance to a nearby star, scientists measure its position in relation to other more distant stars every six months as the earth orbits the sun. The apparent shift of the star when viewed from the two different locations is called *parallax*. In this activity, you will observe parallax on a smaller scale.

1. Cut two pieces of yarn to measure 5 m.
2. Student 1: With your right arm outstretched, hold the end of one piece of yarn directly above your head. Then with your left arm straight down, hold the end of the second piece of yarn against your left leg.
3. Student 2: Hold the loose ends of both pieces of yarn with one hand and stand at a distance where the pieces of yarn are taut, forming a triangle.
4. Student 3: Measure the distance between Student 1 and Student 2. Then using the protractor, measure the angle created by the two ends of yarn in Student 2's hand. Record all data in the table below.
5. Repeat Steps 2–4 by using two shorter or longer pieces of yarn. Record data in the table below.

Length of Yarn (m)	Distance Between Students (m)	Angle of Yarn
5		

ANALYZE AND CONCLUDE:

1. Graph your results below.

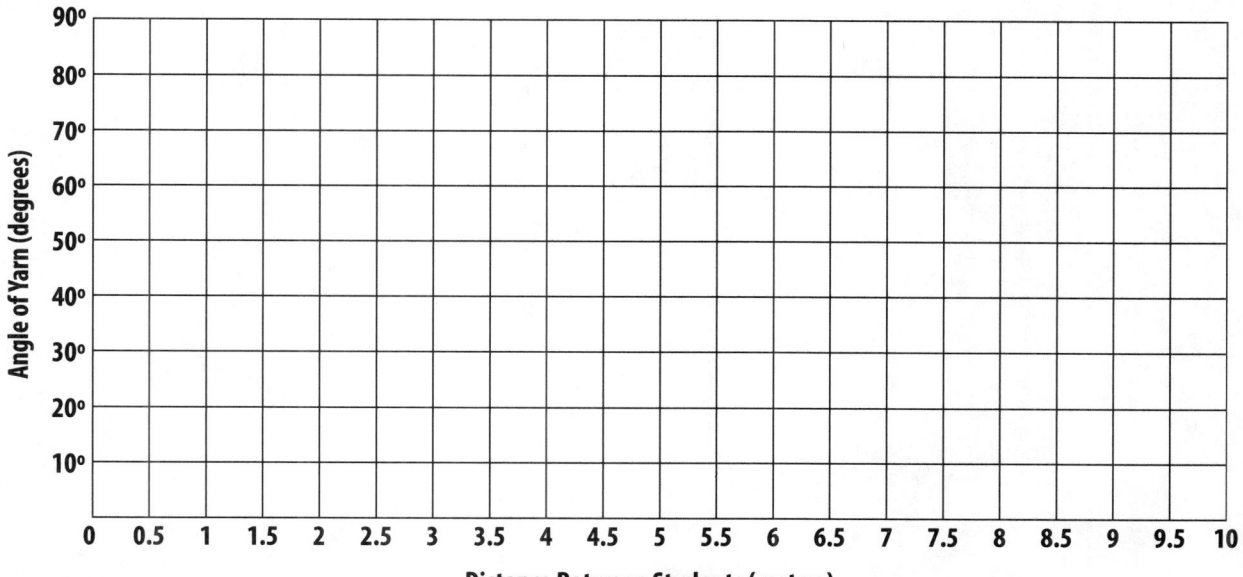

© *Earth and Space Science* • Stars and the Universe

Parallax

WS 8.1.1A

2. How does the distance between students relate to the angle of the yarn? _____

_____
_____
_____
_____

3. How does this activity relate to measuring the distance of a star? _____

_____
_____
_____
_____
_____
_____
_____
_____

Name: _____ Date: _____

Life Cycle of a Star WS 8.1.2A

Label the sequential steps in the life cycle of both a small and a massive star, beginning with a stellar nursery.

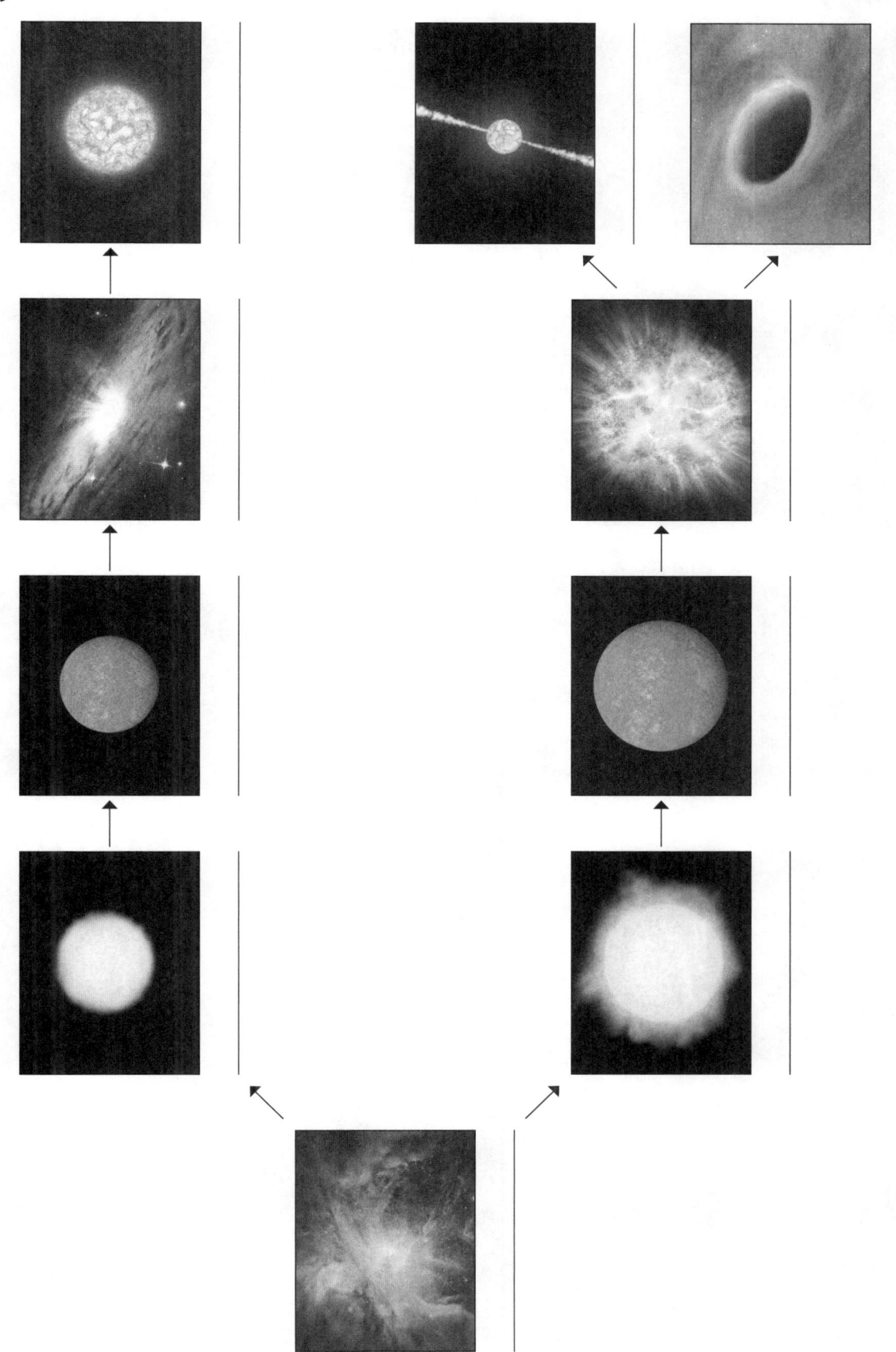

Name: _____ Date: _____

Lab 8.2.1A Rockets

QUESTION: How much cargo can the balloon rocket carry?

HYPOTHESIS: _____

_____

EXPERIMENT:

You will need:	• 2 straight drinking straws	• cargo, such as marbles or paper clips
• fishing line	• 3 large balloons	• mass scale
• scissors	• 2 spring clothespins	• small cardboard box or paper cup
• tape	• meterstick	

Steps:
1. Cut a length of fishing line that will stretch from the ceiling to the floor. Attach the fishing line to the ceiling with tape. Thread the straw onto the line.
2. Attach the other end of the fishing line to the floor. The line should be vertical and taut.
3. Inflate the balloon to the size of a softball. Do not tie it closed; clip it shut with a clothespin.
4. Tape the balloon to the straw, closed end down, and bring the straw down to touch the floor.
5. Release the clothespin and let go of the balloon. Measure how far it travels.

   Record the distance. _____
6. Remove the balloon from the straw and inflate it so it is twice the size of a softball.
7. Repeat Steps 5 and 6. Record the distance. _____
8. Design a rocket to transport cargo. The goal is to transport the heaviest payload possible to the highest height possible. Consider if a multi-stage design is needed. Test and improve your design. You may need to replace your balloons periodically as they lose elasticity. In the Notes section of the table, write your results, any problems you encountered, and what you will change for the next trial.

Trial	Materials Used to Hold Cargo	Cargo and Mass	Distance Traveled	Notes
1				
2				

© Earth and Space Science • Space Exploration

Lab 8.2.1A Rockets

Trial	Materials Used to Hold Cargo	Cargo and Mass	Distance Traveled	Notes
3				
4				
5				
6				

ANALYZE AND CONCLUDE:

1. Explain why the balloon moved. _____
_____
_____
_____

2. What was the maximum payload your rocket could carry? _____

3. What problems did you encounter? _____
_____
_____
_____

4. How did you correct your problems? _____
_____
_____
_____

5. If you could do the experiment again, how would you improve the experiment?
_____
_____
_____

© Earth and Space Science • Space Exploration

Name: _____ Date: _____

Lab 8.2.3A Centripetal Force

MQUESTION: How does centripetal force change when the force of gravity changes?

HYPOTHESIS: _____

_____

EXPERIMENT:

You will need:	• scissors	• permanent marker
• string	• glass tube, 10 cm in length	• stopwatch
• metric ruler	• 5 metal washers	

Steps:
1. Cut a 38-centimeter length of string. Thread the string through the glass tube. If you cannot get the string through, suck gently on the opposite end of the tube. The suction will pull the string through.
2. Label one end of the tube *Top*.
3. Tie 1 washer to the end of the string coming out of the Top.
4. Pull the string through so 15 cm of string is between the end and the washer.
5. Mark the string at the point it comes out of the tube. Be sure to mark all the way around.

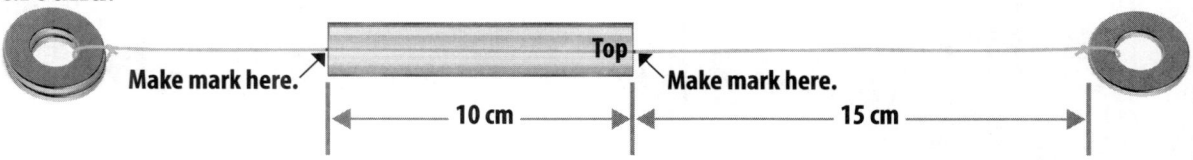

Make mark here. ← 10 cm → Top Make mark here. ← 15 cm →

6. Tie 2 washers to the end of the string coming out the other end of the tube.
7. Mark the string at the point where the string comes out. Be sure to mark the string all the way around.
8. Hold the tube horizontally and spin the single washer. The single washer is the satellite and the two washers at the other end are supplying the gravity.
9. Spin the satellite, keeping the mark on the string right at the opening. This will take some practice to maintain a consistent orbit. Hold only the tube, not the string.
10. When you are ready, have a partner use the stopwatch to measure how long it takes for 20 full circles. Do this five times. Record the results and find the average.
11. Tie 2 more washers to the "gravity" end of the string and repeat Step 10.

	Trial 1	Trial 2	Trial 3	Trial 4	Trial 5	Average
Two Washers						
Four Washers						

© *Earth and Space Science* • Space Exploration

Lab 8.2.3A Centripetal Force

12. For the final step, predict the path the washer will take if the string is cut while the washer is in orbit. _____

13. Spin the washer as you did in Step 11 and then cut the string. Illustrate the path the washer took in relation to the original orbit. Use arrows to show the path of the orbit and the path of the cut washer.

ANALYZE AND CONCLUDE:

1. What provided the gravity in the experiment? _____

What provided the velocity? _____

Name: _____ Date: _____

Lab 8.2.3A Centripetal Force continued

2. What is the name of the force that keeps the washer moving in a circular path?
_____

3. What was the average speed of 20 orbits with two washers? _____
What was the average speed with four washers? _____

4. Why do you think the average speed was different? _____
_____
_____

5. If you were to compare the two-washer and the four-washer tubes to a high Earth orbit and a low Earth orbit, which one would be which? Why?
_____
_____
_____

6. When you cut the string, did the washer follow the path you predicted? _____
_____

7. Explain why the washer did not continue to orbit after the string was cut, but it did continue to move. _____
_____
_____
_____

8. What do you think must happen in order for a satellite to leave Earth's orbit and orbit another celestial body? _____
_____
_____
_____
_____
_____
_____
_____

© Earth and Space Science • Space Exploration

Name: _____   Date: _____

Astronomical Units   WS 8.2.2A

The distance between planets is so large that astronomers use special units to measure the distances. Light travels 9,460,000,000,000 km in one year. This distance is called *a light-year*. Another unit is the astronomical unit (AU). This unit is equal to 149,597,871 km, the average distance between Earth and the sun. Because the universe is so large, the distances between planets and stars are usually given in light-years or AUs. Work the following problems to increase your understanding of the size of the universe. Round answers to three decimal places.

1. Which is longer, a light-year or an AU? _____

2. If a space probe were launched from the earth traveling at 24,000 kph, how many hours would it take to get to the sun? How many days?

3. If a space probe traveled at 35,000 kph, how far could the space probe travel in one year? Convert the answer to AUs.

4. How far could it travel in 20 years? In 68 years?

5. When discussing distances in space, which unit is easier to work with, the kilometer or the light-year? Why?

6. The Crab supernova remnant is about 4,000 light-years away from Earth. Calculate the distance in AUs.

7. What is the distance of 4 light-years?

8. A group of explorers is planning a trip. They start at the moon and travel 5 light-years to Planet A, 8 light-years to Planet B, 2 light-years to Planet C, and then 10 light-years back to the moon. How many light-years did they travel? How many kilometers did they travel?

9. The moon is 363,104 kilometers from the earth. How much closer is the moon than the sun to Earth?

© *Earth and Space Science* • Space Exploration

Name: _____ Date: _____

Orbits
WS 8.2.3A

A satellite's type determines its orbit. Place the following satellite types around the globe: *communications, weather, navigation,* and *scientific*. Label the type of satellite, the type of orbit, and the altitude. Draw arrows showing the orbit direction.

1. Why do you think communications satellites are in their specific orbit? _____
_____
_____
_____

2. Why do you think scientific satellites are in their specific orbit? _____
_____
_____
_____
_____

3. Which type of satellite deals with the most friction? Why? _____
_____
_____
_____

© Earth and Space Science • Space Exploration